AF538423

Powder Metallurgy Equipment Manual

POWDER METALLURGY EQUIPMENT MANUAL

Prepared by the

Powder Metallurgy Equipment Association

Division of

Metal Powder Industries Federation

P.O. Box 2054 Princeton, New Jersey 08540

First Published in 1968, this second edition is completely revised to reflect current practices and the latest process equipment.

Library of Congress Catalog Card No.: 76-523-33

ISBN No. 0-918404-37-1

Printed in the United States of America

UL-2½M-2/77

FOREWORD

Powder metallurgy (P/M) is an exacting technology which requires highly engineered equipment to meet the rigid specifications of the modern powder metallurgist and his customer.

P/M is a capital intensive technology offering economies in materials, labor and energy but exacting a demand on the equipment manufacturer to remain alert to changing needs and advancing industrial sophistication. The manual is published for this purpose.

The manual was prepared by members of the Powder Metallurgy Equipment Association, a division of the Metal Powder Industries Federation. The Federation is an international trade association serving the metal powder producing and consuming industries. Compiled by experts in the field, the manual describes the operation and principal characteristics of the different types of equipment used to consolidate metal powders into shapes and components. It does not attempt to detail the production of P/M products but serves primarily as a practical reference source for individuals already using and familiar with P/M fabricating equipment and the technology of powder metallurgy.

The Manual is divided into four sections. Part One covers the most common types of mechanical and hydraulic compacting systems, including hot forming, coining and sizing presses. Part Two covers tooling and tooling design; Part Three, isostatic compacting and Part Four, sintering. In all cases, the information presented describes commercially available equipment and current production practices.

For information about specific pieces of equipment contact members of the Powder Metallurgy Equipment Association. A directory containing manufacturers' catalogs and names and addresses of leading P/M equipment suppliers is available from PMEA on request. For information about the powder metallurgy process, suppliers of metal powders, P/M parts and other related products, industry standards and publications, contact the Metal Powder Industries Federation. Write to PMEA or the Federation at P.O. Box 2054, Princeton, New Jersey 08540. Telephone (609) 799-3300.

CONTENTS

Page

PART TWO — TOOLING

PART THREE — ISOSTATIC COMPACTING

PART FOUR — SINTERING

1.0 INTRODUCTION TO THE P/M PROCESS

Powder metallurgy is a metal forming process for producing a variety of structural parts and bearings. Powders are blended and then compacted to the required contour in a shaped die. Next, the ejected compact is heated in a controlled atmosphere to bond metallurgically the contacting surfaces of the particles and obtain the desired properties of the part. This Manual covers compacting, tooling, coining and sizing, sintering, and the effects of various gas atmospheres.

Most materials used for compacting are metals in powder form, which range in particle size from minus 80 mesh to minus 325 mesh. Most metals can be compacted to useful shapes by this method. Aluminum, iron, alloy steel, copper, brass, bronze, nickel, and nickel alloys are widely used. However, carbon, graphite, plastics, carbides, ceramics, as well as blends of ceramics and metals, can be shaped by P/M techniques.

Early powder metallurgy parts were bearings and washers—simple shapes with relatively low mechanical properties. Improvements in presses, tooling, powder production, and sintering furnaces have provided the means to produce larger, more intricate and stronger parts. Structural shapes with flanges, hubs, cores, counterbores, and combinations of these, are now commonplace.

P/M components offer many inherent advantages which cannot be attained by any other metalworking processes. For example:

1. Precise Control: The powder metallurgist is able to control his product from the pure powder to the finished part. He can create the material and, at the same time, produce a finished product. Thus, the properties and characteristics of the end product are suited to the demands of the application. Eliminated are poor finishes, uneven internal stresses, impurities, inclusions, unworkable tolerances, or the many other factors which might affect the rate of part production or the quality of the finished product. The manufacturer who uses powder metallurgy parts is assured of uniformity and optimum performance characteristics.

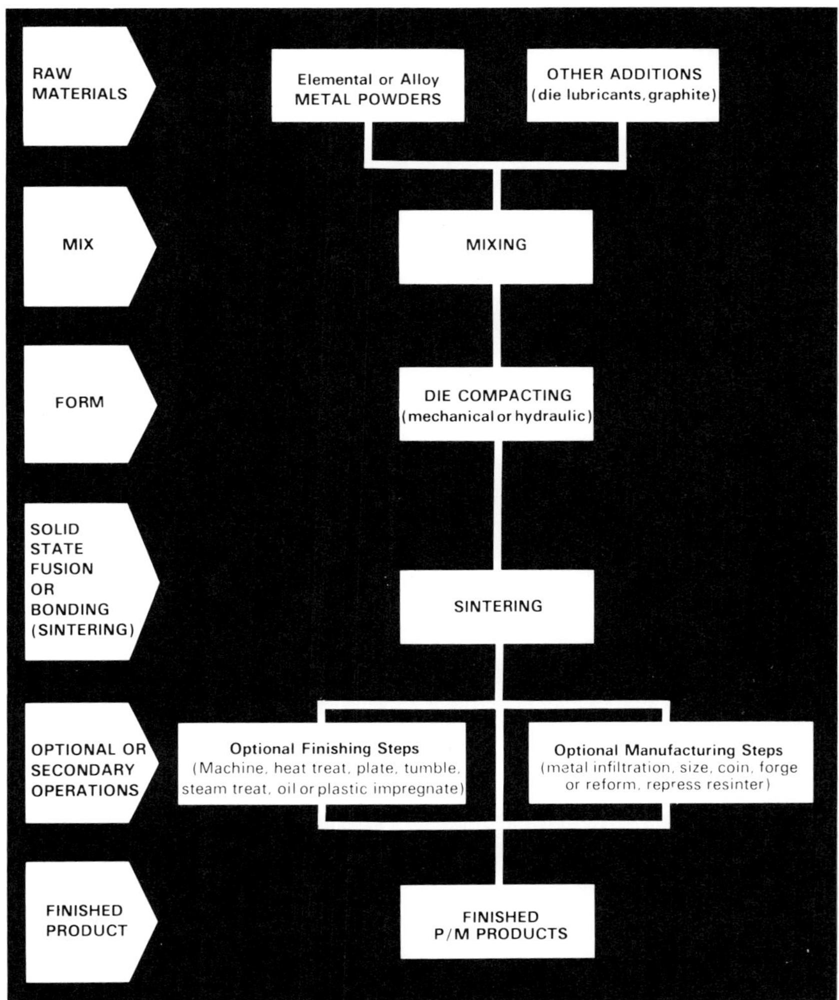

FIGURE 1A.
Powder metallurgy process

2. "Custom-Made" Compositions: Immiscible materials can be combined to produce specific properties not attainable by pyro-metallurgy. Dissimilar metals, non-metallics, plus materials of widely different characteristics can be compacted into parts which have unique properties. Ceramics can be blended with metals. Carbon and copper are compacted to form electrical brushes with wear resistance plus electrical conductivity. Tungsten and silver are immiscible by fusion metallurgy, but their characteristics are combined by powder metallurgy in the manufacture of electrical contacts and switch gear. Copper, tin, iron, lead and graphite are compacted into heavy-duty friction materials. The combining of materials through the use of powder techniques is limited only by research and experimentation.

3. Unusual Physical Properties: Physical properties can be varied from low density, highly porous parts, to high density parts with minimal porosity. Tensile strengths can be varied from low to very high. Dissimilar materials can be compacted in layers to attain a single part with individual properties on opposite surfaces.

Self-lubrication is another example of a unique property attainable only through the use of powder technology. The porosity in a part is controlled so that it can be impregnated with oil or other lubricants to become self-lubricating. Gears and other parts manufactured from metal powder possess self-damping characteristics which permit quieter operations. A powder metallurgy gear does not "ring", yet it can possess all the other features of a wrought metal gear.

4. Reduced Machining And Finishing: The experienced custom parts manufacturer can produce parts in a wide range of finished shapes and sizes which require no further handling and processing. Many gears and cams which require expensive machining to produce from wrought stock can be made from metal powders. Counterbores, flanges, hubs, and holes can be formed when parts are compacted. Keyways and other fastening devices can be made an integral part of the component. It is also common to combine two or more parts into one single unit when product design permits.

5. No Material Loss And Less Material Usage Are Economic Benefits: With the elimination of further processing such as machining and finishing, parts are produced without scrap. The use of powder metallurgy parts reduces inventories of such forms of materials as, bars, rods, plates, sheet and strip.

6. Reproducibility. . . The First Is The Same As The Last: The dies in which powders are compacted are among the most repetitively accurate devices found in any mass production process. Each part made in a die duplicates the preceding part. Part deviation can take place only as the die wears, and with the use of carbide dies this becomes detectable only after tens of thousands of parts are produced.

PART ONE

COMPACTING

2.0 COMPACTING PRESSES

2.1 INTRODUCTION:

The words pressing, compacting, and briquetting are synonymous, and imply cold pressing powders into a green compact. The press acts to change the loose powder to the more densified form, or a green compact. By definition, a press is a machine which applies pressure under controlled conditions of load, direction, stroke, and speed. These controlled conditions are the prime requisites for all compacting presses.

The ultimate objectives in pressing are:

(1) To achieve the required part shape.
(2) To obtain the required green density.
(3) To secure sufficient green strength to permit safe handling of the part.
(4) To provide particle-to-particle contact which is necessary for sintering.

Minimum requirements for any powder metal press should include the following:

(1) Adequate total pressure capability in the direction of pressing, and sufficient part ejection capacity.
(2) Controlled length and speed of compression and ejection strokes.
(3) Adjustable die fill arrangements.
(4) Synchronized timing of press strokes.
(5) Material feed and part removal system.

2.2 TYPES OF COMPACTING PRESSES:

There are three basic types of compacting presses. The first and most widely used is the mechanical press which includes both single punch and rotary type machines. The second type is the hydraulic press. This is only the single punch type. The third major type of press, which is finding increasing usage, during recent years, is the hybrid press. This uses a combination of mechanical, hydraulic, or pneumatic forces to compact a part.

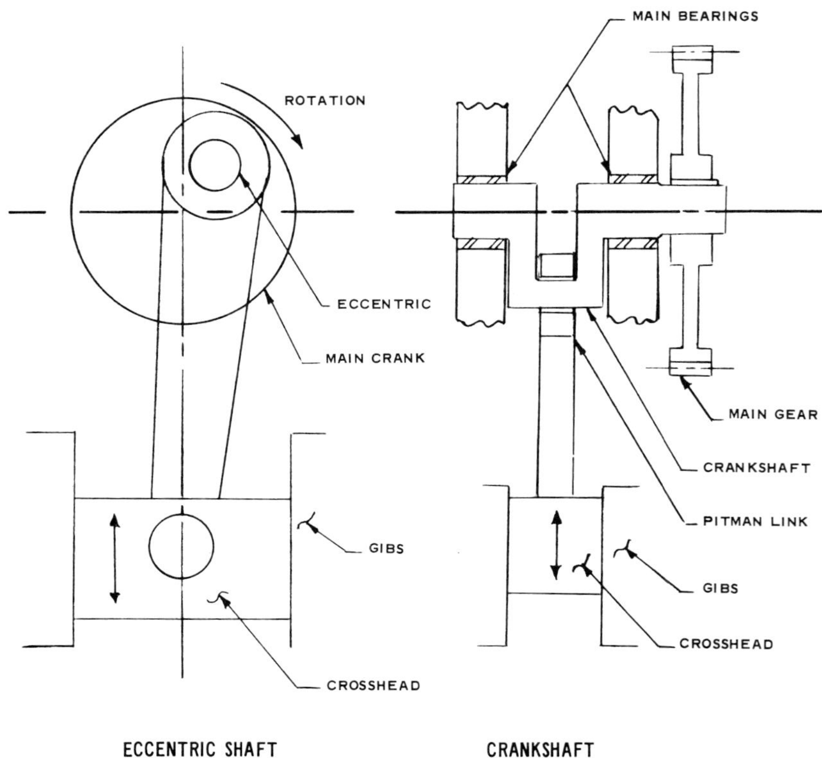

FIGURE 2A.
Eccentric type mechanical press

2.2.1 MECHANICAL PRESSES:

Mechanical presses depend entirely on a predetermined arrangement of mechanical elements for their motion. The arrangement of these elements can be varied, thereby providing the many different types of mechanical presses available.

The most common type of mechanical press is the eccentric or crank type which converts rotary motion to linear motion (see figure 2A). The base mechanism permits high loading with low torque requirements at the maximum compression point (bottom dead center) with a low final pressing speed. Pressure in this type of unit is normally applied only from the top with an adjustment on the crank, eccentric, or pitman arm to control the amount of linear slide or crosshead travel.

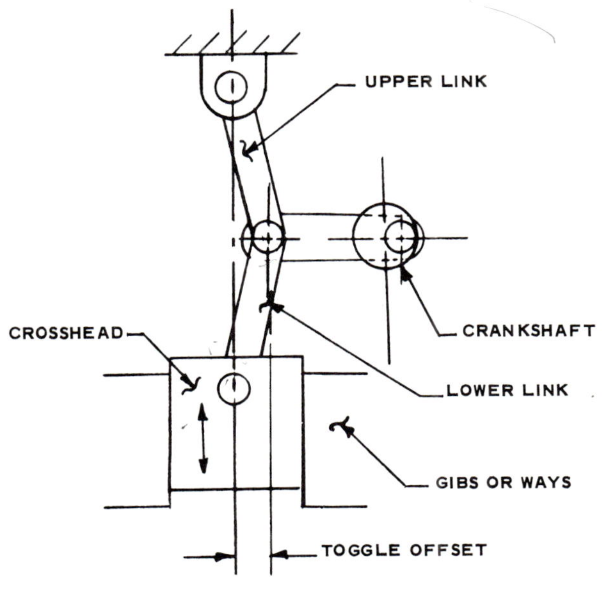

FIGURE 2B.
Toggle mechanism

The second most common type of mechanical press is the toggle or knuckle type. Actuation is usually accomplished by an eccentric which straightens a jointed arm or lever, the upper end of which is fixed at the top while the lower is guided for controlled accurate motion. The toggle design enables the press to develop very high pressure at low power and also to provide the minimum speed near the end of the compacting stroke. This minimum speed provides a natural type dwell for the compression point of the press cycle (see figure 2B).

The third type of mechanical press is the cam type. These presses utilize cams with a mating lever arrangement to convert rotary motion to linear motion. Pressing speed, timing, and motion are controlled by changing the size and shape of the cams (see figure 2C).

The last type of mechanical press which will be mentioned is the rotary press. This machine has a series of punches and dies arranged in a common tool holding member, the press head or turret, which rotates around a spindle or stand. This provides a fixed reference point for mounting the press cams. The cams provide linear motion to both the top and bottom punches. Rotary presses are best suited for high volume, high production applications. They are generally available in tonnage ranges from four to one hundred tons. Production rates can range from a few pieces to several hundred parts per minute depending upon the material flow, part configuration, part dimension, and pressing tonnage.

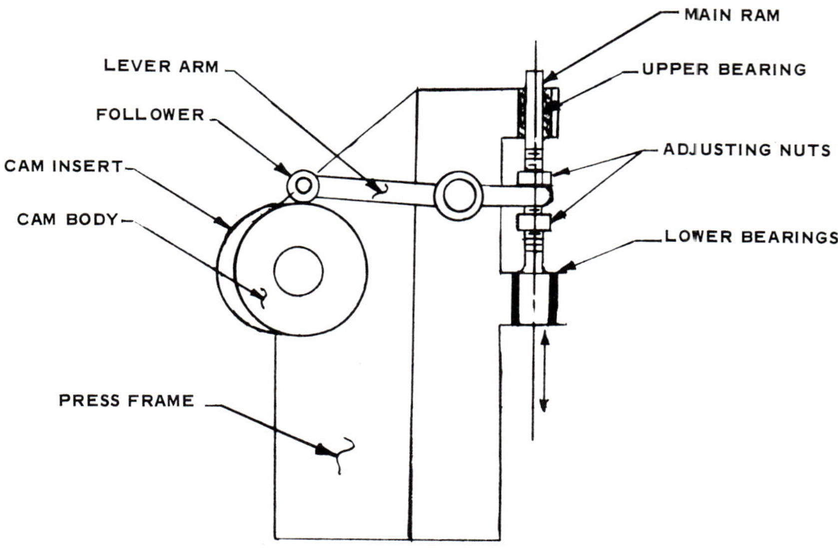

FIGURE 2C.
Cam action

In general, mechanical presses, whether single punch or rotary, use flywheel energy during the pressing and ejection portion of the cycle. Because the application of the load on the press during the compacting and ejection portion of the cycle can be considered to be intermittent, a flywheel is used to maintain uniform rotation on the press drive members. Drive motor horsepower can thus be kept low.

2.2.2 HYDRAULIC PRESSES:

Hydraulic presses are used for the production of all classes of parts. Capacities range from 10 tons to 3000 tons and are commercially available from several machine builders. Up till this time, the only exception to the application of hydraulic power has been the rotary press. However, even rotary presses may use hydraulic control elements.

Types of hydraulic presses are the single action, floating die, double action, multiple platen, and the floating die/withdrawal die-set types. The die-set press is available with or without removable die-set features.

3.0 FUNDAMENTAL REQUIREMENTS FOR SELECTION OF PRESSES

Tonnage capacity and stroke are important considerations in selecting a pressing system. All press manufacturers list rated tonnage capacities and stroking dimensions in their specification data sheets. The tonnage listing describes the maximum force the press will be capable of exerting during compression while maintaining acceptable life of the press components. Stroke is usually given in terms of punch travel while capacity is stated as depth of die fill and ejection stroke in inches.

With adequate tonnage and stroke when pressing a multi-level part, the press should support each separate level of the part as an independent tooling member. The ultimate objective in supporting all sections of the part will be to maintain density control in the different sections of the part. This will provide uniformity at all levels.

3.1 TONNAGE CAPACITY:

The total load requirement of a press in tons is a function of the molding pressure of the material multiplied by the projected area of the part to be compacted. The molding pressure requirements are determined from the final density data on the part in relationship to the particular material properties one desires. See tonnage requirements and compression ratios for various powder products. This type of data is normally furnished by all powder producers for their various products.

In practice if the example of a material requiring 50 tons/in ^{2} (7 ton/cm ^{2}) and a part with a projected area of one square inch or (6.45 cm^2) is studied.

$$\text{Press load (tons)} = \text{Molding Pressure lbs/in}^2 \times \text{Projected Area in}^2$$

$$50 \text{ tons} = 50 \text{ tons/in.}^2 \times 1 \text{ in.}^2$$

Dimensional analysis of the above simple equation can be used as a cross check in calculations as tons must be the final unit on both sides of the equal sign.

Our example dictates that the minimum press size for this part application must be at least 50 tons capacity.

3.2 EJECTION CAPACITY:

Ejection capacity is an important rating of a press. It is usually stated as tons by the manufacturer of the press. Some press manufacturers, because of the design of their machines, divide rated ejection loading. Initially they will list ejection capacity equivalent to compressive loads for a pre-defined ejection breakaway stroke, which is usually between 1/32 in. (0.079 cm) and ½ in. (1.27 cm) of the start of the stroke. A reduced sustained ejection load, normally between 25% to 50% of the maximum compressive load rating, will also be listed. Ejection load depends on punch, core rod and die side wall contact areas, tooling material, tool surface finish, and amount and type of lubricants. For a rough calculation of ejection load, multiply the part contact area with the tooling in the axial direction of pressing by the ejection breakaway pressure of the material.

3.3 STROKE CAPACITY:

The second consideration in selecting a press for a specific part is to determine if the press has sufficient die fill and ejection stroke. During compaction in the die, the volume of the filled powder is reduced. The ratio of the volume of loose powder (apparent density) to the volume of the compacted part (green density) is the compression ratio of the powder (fig. 3A).

Green density is the density of the part after it has been pressed but before it has been put through the next process — sintering in the case of metal powders, or firing in the case of ceramics or ferrites. The most common way of expressing density is grams per cubic centimeter. (See MPIF Standards 4, 28, 35, and 42 for additional details.)

FIGURE 3A.

In practice, if a part of 2 in. (51 mm)thickness is considered for production to a green density of 6.25 gms/cc, and if the powder selected has a compression ratio of 2.5:1 to secure this density, then a 5 inch or 127 mm die fill is needed for the part. Figure 3B illustrates a straight wall die, lower punch, and upper punch, with loose powder in the die cavity in the fill position. The distance from the top surface of the die to the top surface of the lower punch is defined as the depth of fill. The amount of powder fill to produce a part is determined by multiplying the finished part thickness by the compression ratio of material to be compacted to the required green density.

Ejection stroke is measured as the distance from the top of the lower punch to the top of the die in the compression or press position (fig. 3B, compression position). The press at this stage must have enough stroking capability to either push the part up to the top surface of the die or to strip the die down to a point where the top surface is flush with the lower punch, so the part can be removed. The normal ejection stroke is equivalent to the thickness of the part plus the total amount of upper punch penetration into the die.

The amount of force required to obtain a given green density depends upon the material being pressed. It can range from as low as 3 tons per square inch

(4.23 kg/mm^2) to as high as 60 tons per square inch (84.6 kg/mm^2) of surface area. The upper limit is usually held to 60 tons/in.^{2} to provide a safety factor against premature tool failure under load. Table 3.1 gives some typical examples.

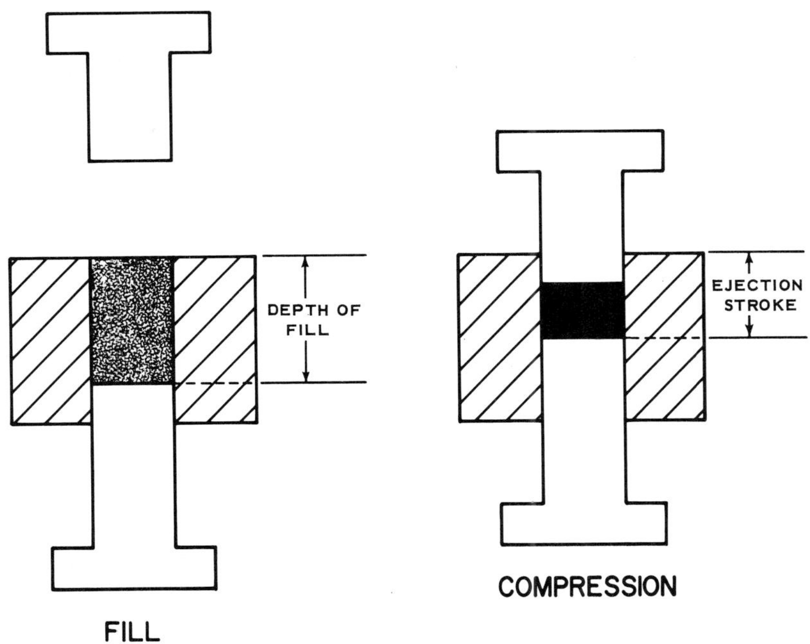

FIGURE 3B.

3.1 TABLE

3.4 TONNAGE REQUIREMENTS AND COMPRESSION RATIOS FOR VARIOUS MATERIALS

Type of Material or Part	Tons per Sq. Inch	Compression Ratio
Aluminum	5 to 20	1.5 to 1.9:1
Brass	30 to 50	2.4 to 2.6:1
Bronze	15 to 20	2.5 to 2.7:1
Carbon	10 to 12	3.0:1
Copper-Graphite brushes	25 to 30	2.0 to 3.0:1
Carbides	10 to 30	2.0 to 3.0:1
Alumina	8 to 10	2.5:1
Steatites	3 to 5	2.8:1
Ferrites	8 to 12	3.0:1
Iron Bearings	15 to 25	2.2:1
Iron Parts:		
low density	25 to 30	2.0 to 2.4:1
medium density	30 to 40	2.1 to 2.5:1
high density	35 to 60	2.4 to 2.8:1
Iron Powder cores	10 to 50	1.5 to 3.5:1
Tungsten	5 to 10	2.5:1
Tantalum	5 to 10	2.5:1

The above tonnage requirements and compression ratios are only approximations and will vary with changes in chemical, metallurgical, and sieve characteristics; with the amount of binder or die lubricants used; and with mixing procedures.

(Note: To convert tons per sq. in. to kg per mm^2, multiply tons by 1.41)

3.5 NEUTRAL AXIS:

When compaction begins, as the upper punch enters the die and compresses the powder, the density of the material normally increases first at the face of the upper punch. Friction develops from four sources: (1) between the partially compacted powder and sidewalls of the die, (2) between the individual particles of the powder repacking to reduce voids in the early stages of compacting, (3) between particle to particle from plastic flow as the particles are distorted in the later stages of compaction, and (4) between the punch faces and powder particles due to sideways movement of the grains.

Since powders do not exhibit hydrostatic flow, the friction generated absorbs part of the force being applied by the upper punch. This loss of force is not important when pressing thin parts. When thicker parts are pressed, force must be applied to both the top and bottom of the part to compensate for some of the frictional loss. The ability to transmit force throughout the powder being compacted determines the uniformity of density of the green compact. Uniform density is important to insure dimensional control during sintering. Generally, uniform density is not reached when the part length/diameter ratio exceeds 3:1.

Under pressure, powders will not flow from one part level to another. Therefore, when parts of more than one level are being pressed, a separate pressing force must be supplied for each level. The thickness of each level will determine whether the pressing force must be applied from one direction or from both directions. On parts with more than one level, the pressing forces must be applied to all levels simultaneously to assure uniformity.

The neutral axis is the low density plane in the green compact that is perpendicular to the direction of pressing.

The plane is present in the part because the powder tends to resist flowing under pressure, which results in a lesser densification of the material farthest away from the forming element of the tooling. Placement of this plane in the part is controlled by the relative tooling motions. Normal positional location is usually set at the center of the part. When there is more than one level being compacted, there will be a neutral axis present for each part level (fig. 3C).

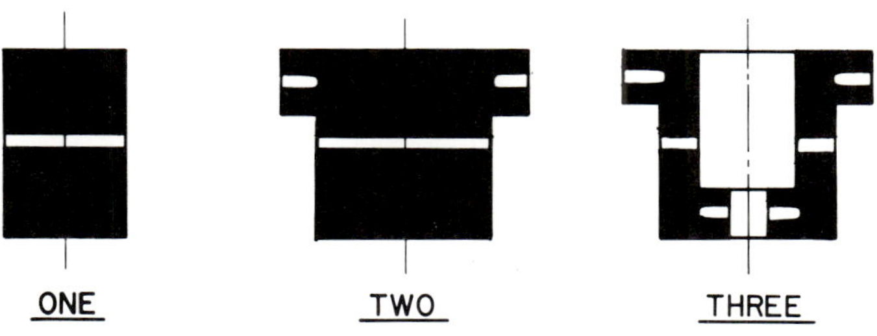

FIGURE 3C.
Neutral Axis

4.0 PRESSES AND TOOLING SYSTEMS AS RELATED TO PART CLASSIFICATIONS

P/M parts usually are classified by evaluating the complexity of part design. The general scale is based on a classifying range of I through IV, as shown by figures 4A to 4D. The more complex parts, as shown in figure 4D, require more complex press and tooling systems to achieve uniform density throughout all levels of the part.

4.1 GENERAL PART CLASSIFICATIONS:

Class I parts can be defined as thin, one-level parts of any contour that can be pressed with a force from one direction. The A dimension, as shown in the schematic section of figure 4A, is generally limited to a maximum of 0.250 in. (0.635 cm).

Class II parts, illustrated in figure 4B, are also one-level parts of any thickness and contour that must be pressed with forces from two directions.

Class III parts, as shown in figure 4C, are two-level parts of any thickness and contour that must be pressed with forces from two directions.

Class IV parts, figure 4D, can be defined as multi-level parts of any thickness and contour that must be pressed with forces from two directions.

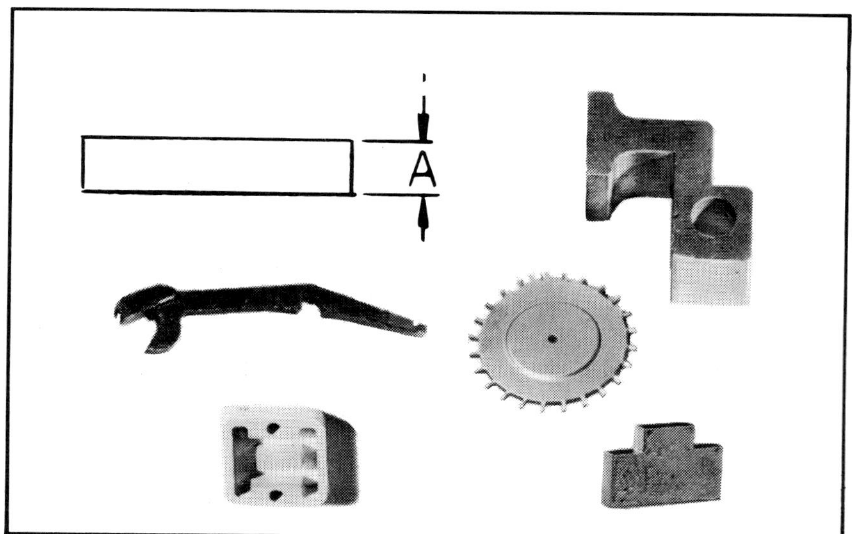

FIGURE 4A.
Class I Parts

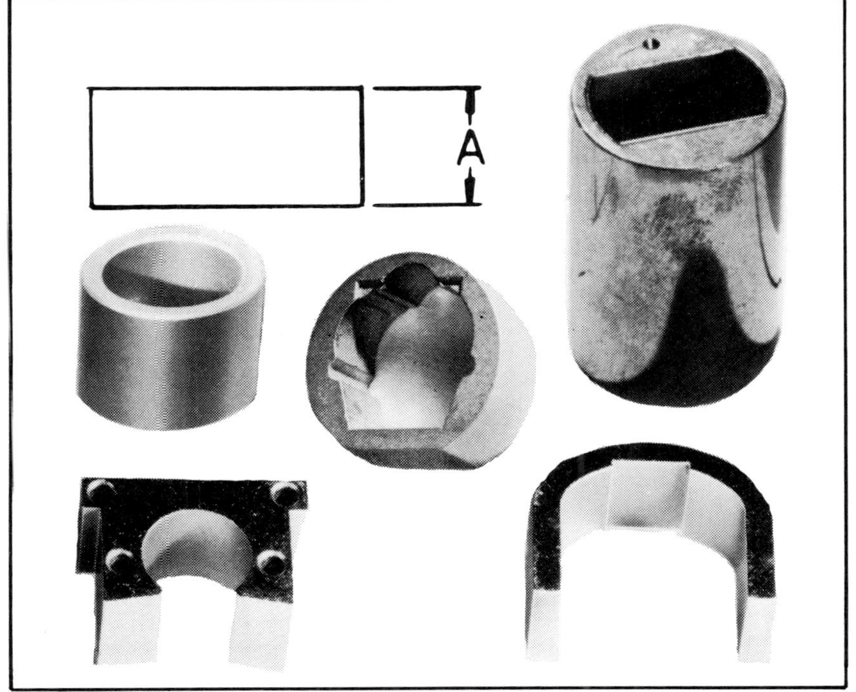

FIGURE 4B.
Class II Parts

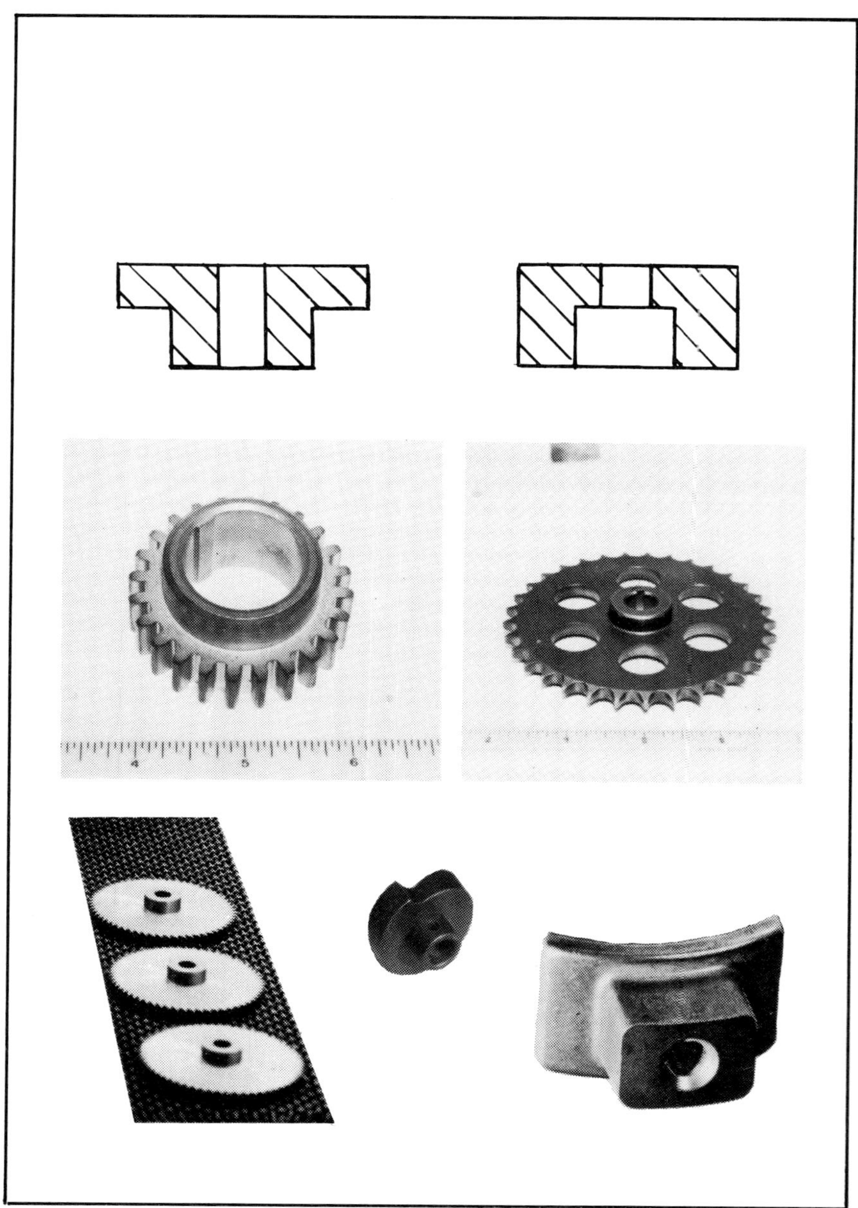

FIGURE 4C.
Class III Parts

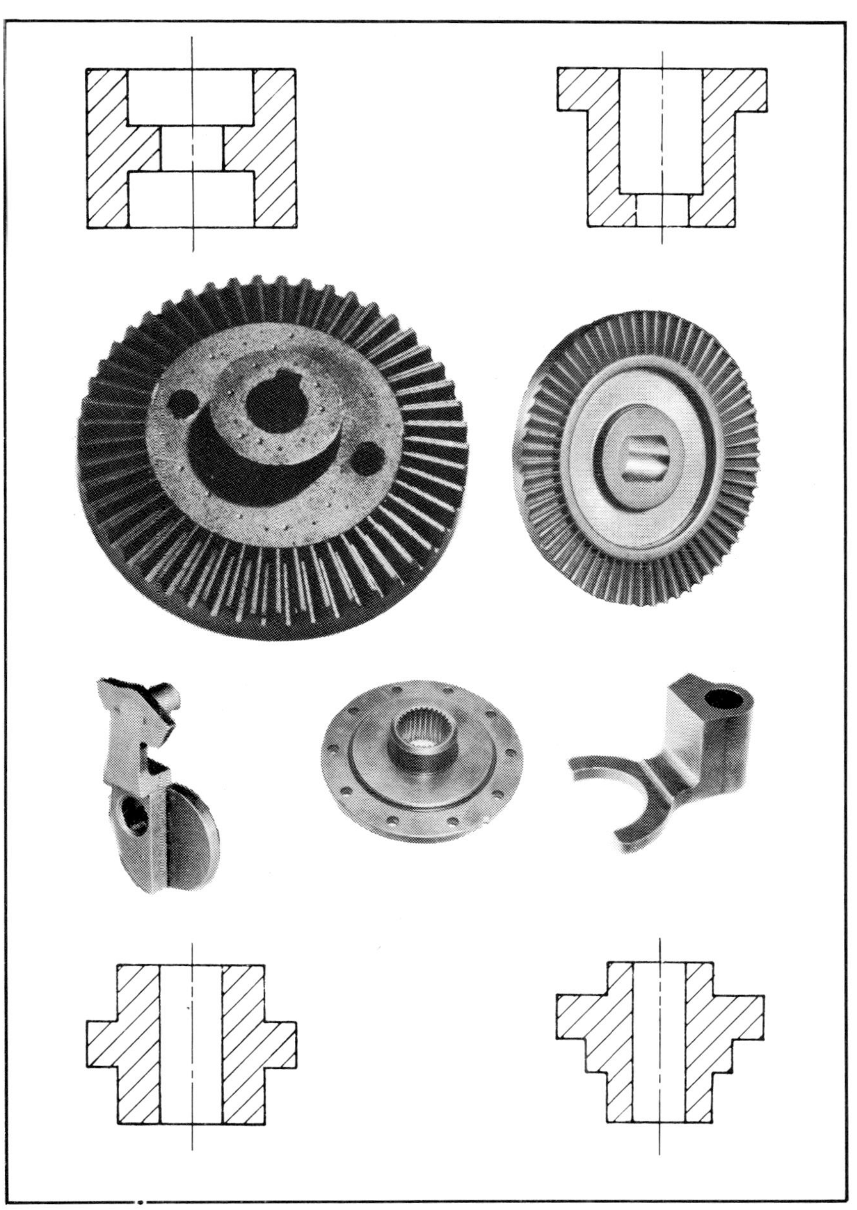

FIGURE 4D.
Class IV Parts

4.2 DIMENSIONAL TOLERANCES IN COMPACTING:

The economics of producing parts by P/M, much like the economics in any other metalworking process, are to a large extent governed by the dimensional tolerances specified. Ordinarily tolerances should be as liberal as possible to give maximum tool life during compacting. If the required tolerances can be held by compacting and sintering only, secondary sizing, which adds to costs, is unnecessary.

In general, the following practical tolerances can be normally achieved on parts in the as-formed and as-sintered condition:*

* Recommended tolerances are contained in MPIF Standard No. 44.

A. Length - in direction of pressing:
   1 in. (2.54 cm) or less– ±0.002 (0.0051 cm)
   1 in. (2.54 cm) to 3 in. (7.62 cm)— ±0.005 (0.013 cm)
   Over 3 in. (7.62 cm) ±0.010 (0.025 cm)
B. Radial - right angles to direction of pressing:
   ±0.001 to 0.002 in./in. (0.0025 to 0.005 cm)
C. Cored holes:
   ±0.0005 in. (0.0013 cm) to ±0.001 in. (0.0025 cm)
D. Concentricity:
   Up to 1 in. (2.54 cm) bore - 0.002 in. (0.005 cm) T.I.R.
   1 in. (2.54 cm) to 2 in. (5.08 cm) - 0.003 in. (0.0076 cm) T.I.R.
   2 in. (5.08 cm) to 3 in. (7.62 cm) - 0.004 (0.010 cm) T.I.R.
   More than 3 in. (7.62 cm) - 0.006 in. (0.015 cm) T.I.R.
E. Gear tolerances are usually based on the A.G.M.A. classification numbers. In normal processing powder metallurgy gears can be held to A.G.M.A. quality numbers 5 to 7. By utilizing special sizing techniques, gears of A.G.M.A. quality No. 8 can be produced.
F. Sizing after sintering will give closer tolerances, but at added cost. For example, length tolerances in the direction of pressing can be held to ±0.001 in. (0.0025 cm). Hole diameter tolerancing can usually be held to ±0.0005 in. (0.00125 cm).

4.3 SINGLE ACTION PRESSES:

Single Action Opposed Ram: A single action opposed ram type press (or tooling system) includes a die to form the outer contour of the part, an upper punch to form the top surface of the part, and a lower punch to form the bottom surface of the part. It can also include core rods to form any through holes.

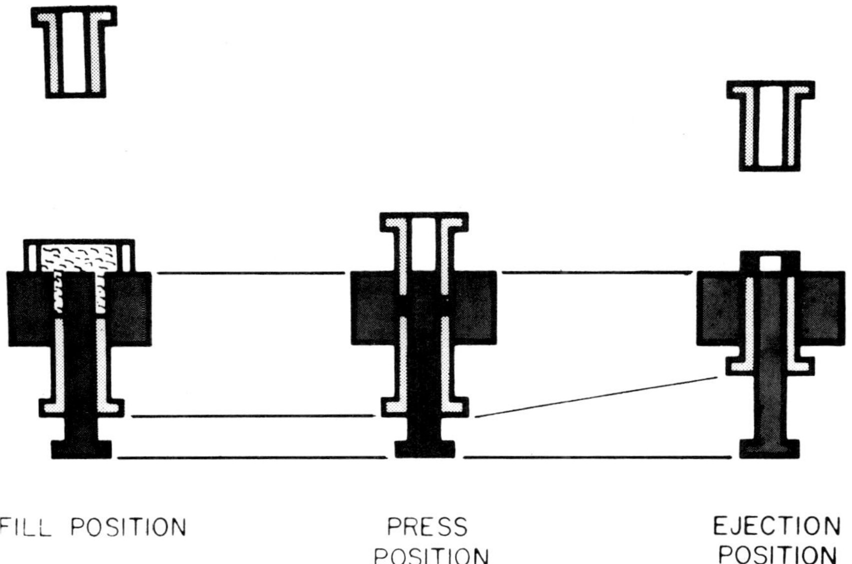

FIGURE 4E.
Single action press system

During the compaction portion of the pressing cycle, the die, lower punch, and core rod, if any, remain stationary. In operation the upper punch enters the die and compresses the powder against the stationary lower punch and inner surface of the die (see fig. 4E). Because the force applied by the press ram is from one direction only, it is called a single action press. Ejection of the compacted part can be accomplished in either of two ways: the die and core rod remain stationary while the lower punch raises the part from the die cavity, or the lower punch remains stationary while the die and core rod are lowered to a point where the die table surface is flush with the top of the stationary punch.

Single action compacting presses may be either mechanical or hydraulic. The single action mechanical press has the highest speed. However, it generally is used to produce only the thinner Class I type parts, particularly if uniform part density and good dimensional tolerances are required.

Single Action Anvil: A specialized form of single action press is the anvil type compacting system. It includes a die to form the part contour, and a lower punch to form the bottom surface of the part. An upper punch is not used, since the top section of the part is formed by an anvil or flat surface which slides over the filled die to level the powder; any loose powder is pushed away. The anvil is then clamped to the die at the compression position.

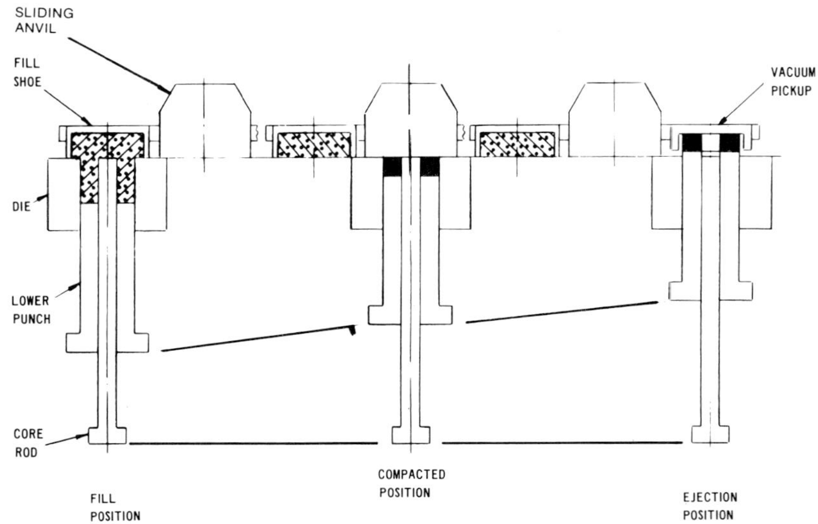

FIGURE 4F.
Sliding anvil press system

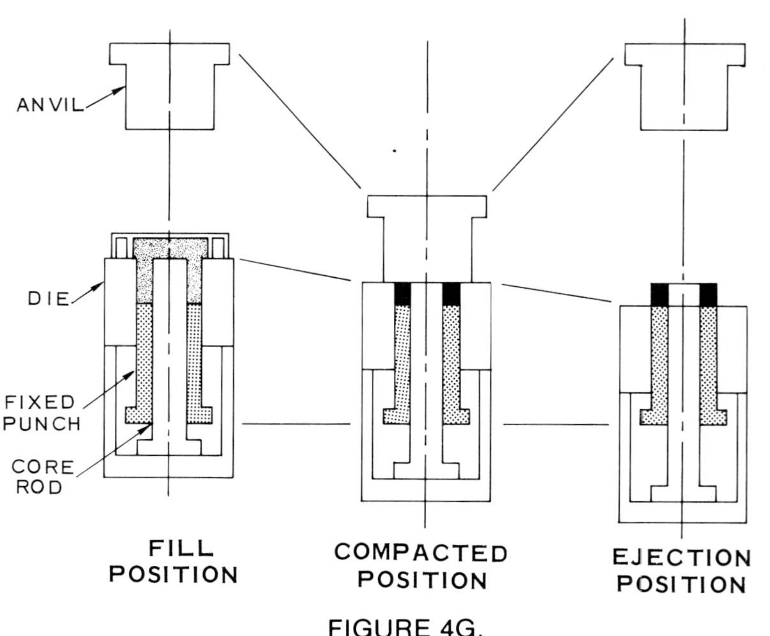

FIGURE 4G.
Anvil withdrawal system

Compaction can be by either of two methods of press action. One utilizes movement of the lower punch for fill, compression, and ejection of the part, while the die table is used as a stationary reference (see fig. 4F). The other action utilizes the lower punch as the fixed reference, while the die table is moved into position for fill, compression, and ejection during the press cycle (fig. 4G). Both actions can be defined as a single level pressing, since the compaction force is applied only from one direction. They are limited in terms of part thickness related to uniform density, and normally are used for the production of Class I type parts. This system does offer the advantage of one less tooling member (upper punch), thereby minimizing the tool set costs and set-up time required for upper punches. Multiple cavity tooling with this system is easily implemented, providing additional machine production capacity.

4.4 DOUBLE ACTION PRESSES:

Double Action Opposed Ram: The double action opposed ram press, or tooling system, applies an equal force to the powder simultaneously from opposite directions through movement of the upper and lower press rams (fig. 4H). The lower punch normally has three separate adjustment positions: powder fill, compression, and ejection height. The upper punch is also independently adjustable to control the depth of penetration into the die. These adjustments permit positive control to position the neutral axis properly. Stationary and movable core rods can be used when parts with through or blind holes are being produced.

The system is used primarily for the production of Class I and Class II type parts; however, some Class III parts are suited to these motions. It will produce parts with a more uniform density than the single action press system.

Opposed ram double action type presses are available with either hydraulic or mechanical drives.

Double Action Floating Die: The double action floating die press or tooling system normally consists of a moving upper punch, moving die table or die element, and a stationary lower punch. The die table or die element is held in position during the fill portion of the pressing cycle by a spring, air cylinder, or hydraulic cylinder. During early compacting the upper punch moves into the die cavity and, as noted before, friction is developed between the powder and the inside surface of the die*. When the frictional forces overcome the die supporting or counterbalancing forces, the die descends at the same speed of the upper punch. This die movement has the same effect as a movable lower punch, thus applying pressure on the powder from both top and bottom. (fig. 4J) Normally a part pressed by this system will have density variations determined by the supporting force of the die. The neutral axis may not be located at the center of the part.

*See section 3.5, p. 12

In order to overcome the density variation, an outside force, equal to the die support force, is applied to the top surface of the die as the upper punch enters the die cavity. With the application of the equalizing force, the die will move downward at a rate equivalent equal to ½ the speed of the decending upper punch. The resulting uniform pressure from both directions produces a part with the neutral axis in the center.

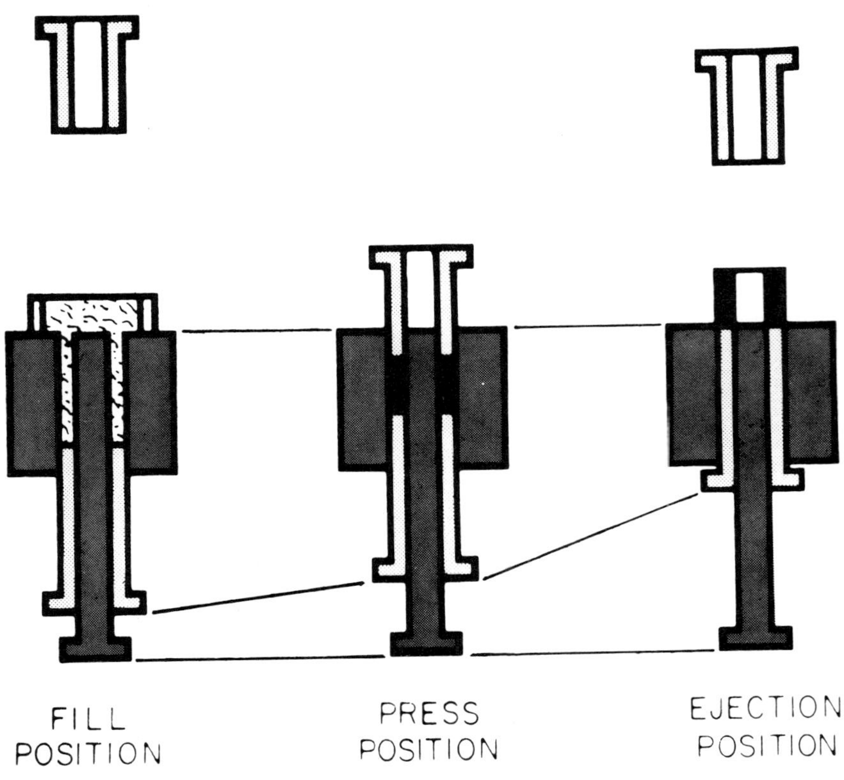

FIGURE 4H.
Double action press system

After compaction, the die remains in contact with the part, while the lower punch moves up to push the part out of the die cavity for ejection. Systems of this type are used mainly for the production of Class I or Class II parts. With modifications in tooling and the addition of positive stops on the die table or die elements for the compression position, parts of Class III and Class IV types can be produced.

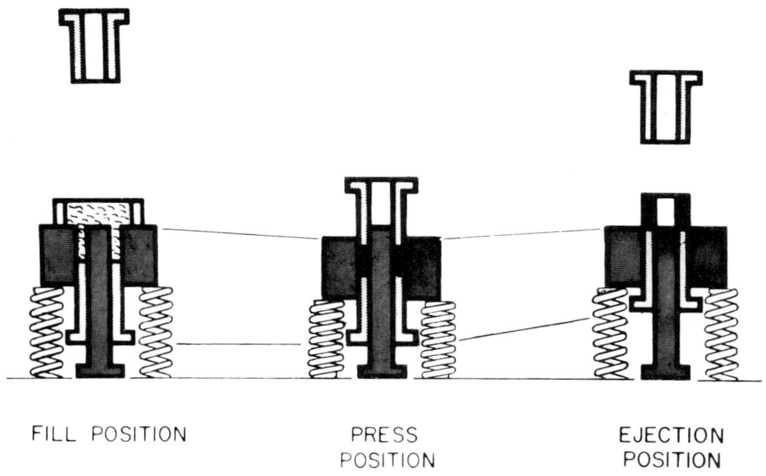

FIGURE 4J.
Floating die press cycle top & bottom pressure

4.5 WITHDRAWAL PRESS SYSTEMS:

This pressing system utilizes downward movement of the die into the compacted or press position, and a further downward movement of the die for ejection. The lower punch is a stationary member and acts as a reference level for fill, press, and ejection, since the die must be withdrawn to a point flush with the top of this punch for full part ejection. At this stage the part is completely supported by the fixed punch. The die is adjustable in three positions: (1) filling position, (2) compacted position, and (3) ejection position.

Withdrawal Controlled Die Motion: The withdrawal controlled die motion press system is based on a double action non-simultaneous compacting method. The lower punch remains stationary and is utilized as the dimensional reference tool for fill, press, and ejection (fig. 4K). The die table is adjustable in three positions: (1) fill, (2) press or compacted position, and (3) ejection. The timing of the die movement relative to the upper punch movement can be controlled through the compacting cycle. Adjustment of timing on the die travel provides positive control over the position of the part's neutral axis.

To generate the non-simultaneous double action movement:

(1) the upper punch moves down and seals off the die cavity. The upper punch and die now travel at the same speed, relative to the fixed lower punch, until the die controlling element reaches a positive mechanical stop. This movement sets up as a single action compacting mode over the fixed punch and will place the neutral axis at this time at a point slightly above the normal center of the part.

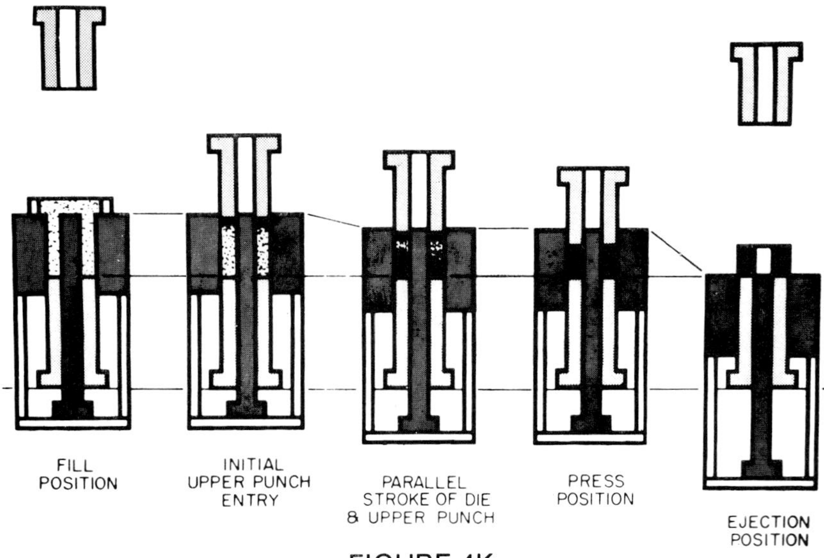

FIGURE 4K.
Withdrawal press system

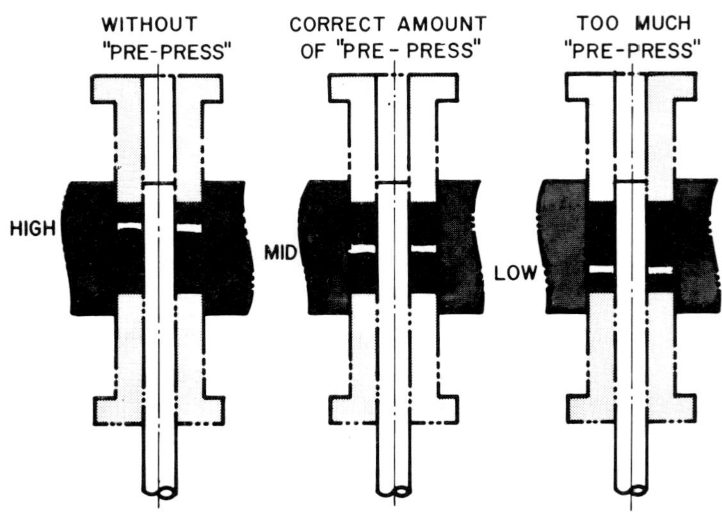

FIGURE 4L.
Positive control of "neutral axis"

(2) prior to completion of compacting, the die is held stationary against its stop, while the upper punch moves an additional amount to penetrate the die further. This final movement, which can be considered single action pressing from the top, places the neutral axis back in the center of the part. Figure 4L illustrates the positive control of the neutral axis through the final upper punch penetration or pre-press adjustment

Withdrawal Floating Die Press System: The floating die withdrawal system is a compacting mode which utilizes a free float on the die table into the press or compression position. It utilizes mechanical or hydraulic control to move the die table into the ejection position. The compacting phase of this cycle is very similar to the double action floating die system. Parts are ejected by the downward movement of the die element or table over the fixed lower punch. The function of this fixed lower punch is the same as for the controlled die motion system: it establishes a reference level for the die table in the fill, press, and ejection positions. Most systems of this type have mechanical stops to assure accurate positioning of the die with respect to the fixed punch at the compacted or press position. Figure 4M schematically illustrates this type of system.

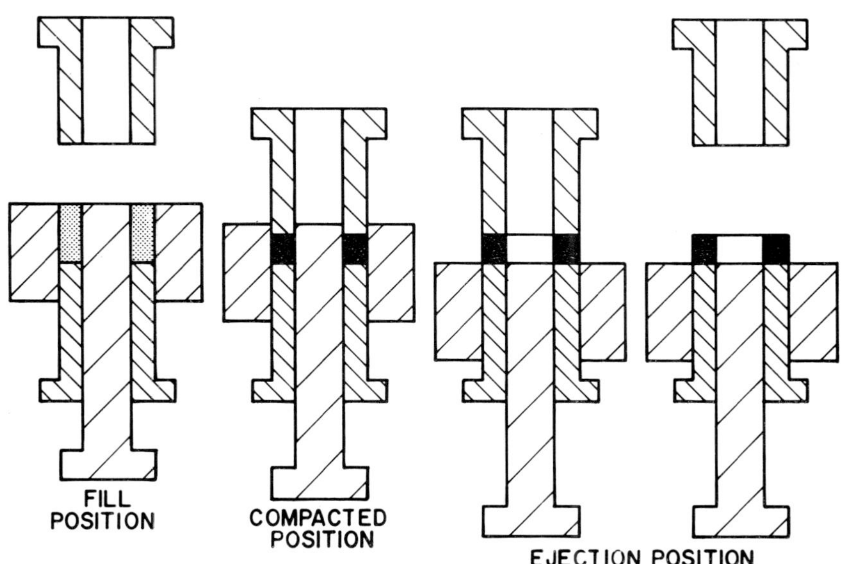

FIGURE 4M.
Withdrawal-floating die-single punch

4.6 MULTIPLE MOTION PRESSES:

Multiple motion presses, or tooling systems, are those which, on multi-level parts, support each level of the part with a separate punch or tool member. This type of system assures minimum density gradients in all levels of the part.

For example, if an external flanged part were compacted in a die with a fixed shoulder, the flange and body would have different green densities. In effect this would be single action pressing on the flange and double action pressing on the body. Usually the result is a crack on the underside of the flange where it joins the body. A further fault is that during sintering there would be differential movement (growth or shrinkage) between the flange and body, leading to warpage and loss of dimensional control. At ejection, regardless of the type of ejection motion (die withdrawal or punch push-out), the part is unsupported on the flange section and further cracking can develop. Utilizing the same approach with an internal flange or counter-bore configuration with a stationary type shouldered core rod also produces the same results.

The multiple motion double action pressing system eliminates crack formation problems. Each lower part level has been tooled with a separate punch. The distance at which the lower punches move relative to one another is normally adjustable, allowing a uniform build-up of density to take place for each level of the part. The tooling members normally have independent adjustments for powder fill, compression, and ejection. To produce double hub parts, the timing of the secondary punch motions can be adjusted to transfer powder into the different upper punch levels.

Multiple motion presses or tooling systems are available in the dual action-opposed ram, withdrawal, controlled die or floating die styles of presses, and in anvil presses where the part configuration has one flat side. The systems can be used to produce parts of all four general classes.

Multiple Motion Presses-Opposed Ram: The opposed ram multiple motion press or tool system is similar in action to the double action opposed ram system. On mechanical presses, the main lower motion is transmitted by cams through linkages to the primary lower punch, as well as the secondary lower punch. However, additional tool elements can be positioned by hydraulics or pneumatics. On all systems of this type, the die platen is held fixed. Ejection takes place by upward movement of the lower punches and is usually sequenced so that the part is completely supported throughout all ejection phases.

Figure 4N illustrates a typical system compacting a complex Class IV shape. The configuration required powder transfer from the lower punches into the upper punch hub volume before compaction begins. Powder is transferred by adjusting the upper inner and outer punches to provide an adequate volume of powder for the proper density in the hub section. The upper outer punch will enter the die ahead of the upper inner punch, and this will result in a loss of powder unless provisions are made to confine the

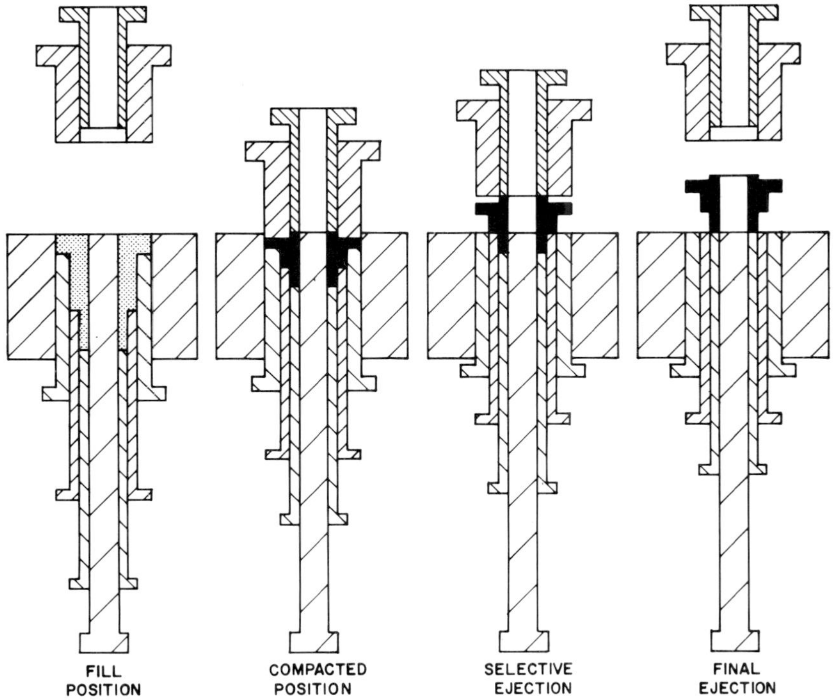

FIGURE 4N.
Stationary die-multiple punches

powder to the designated volumetric cavities. Usually this is done by positioning a moving core pin (mechanical, pneumatic, or hydraulic) in the upper inner punch to seal off the center section of the part for transfer.

Final compacting pressure takes place after powder transfer by all lower punches moving up and all upper punches moving down. The center core in the upper punch floats back and the die remains stationary. Ejection occurs as all punches move up, supporting each level of the part until all levels have been ejected from the die.

Multiple Motion Floating Die Withdrawal: In the floating die multiple-motion press the lower motions are obtained by floating or counterbalancing the lower punches with air, oil or spring cushions. Punches are usually mounted on individual platens or rams which have independent adjustments for fill, press and ejection positions. The punch forming the lowest surface of the part does not move and is considered as a stationary ejection reference. The die platen must be driven down to a point flush with the top surface of this punch for full part ejection.

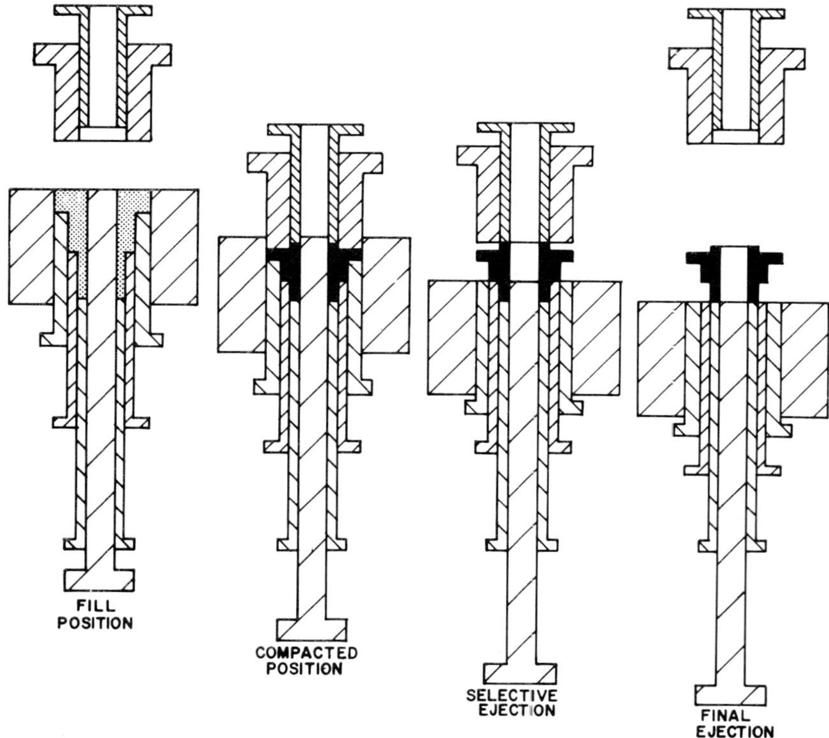

FIGURE 4O.
Withdrawal floating die multiple punches

Figure 4O shows a typical Class IV part as compacted by this system. All moving tool elements travel downward during the compaction portion of the cycle to positive stops. Before ejection, these stops are released so the lower platens or rams can be moved further down. During ejection the upper punches move off the part, while the lower punches, in a sequential manner, move downward with the die mantle movement until all tool members are level with the top of the stationary or fixed punch.

By timing the die descent in relation to the upper punch movement and through control of the counter-balance forces on the tooling platens or rams, the neutral zone of the part can be controlled.

Multiple Motion Controlled Die Withdrawal: In the controlled die movement multiple motion withdrawal system, all moving tooling elements travel downward through the compaction cycle as with the floating die system. However, this system normally utilizes the removable tool holder or die-set concept of tooling such as shown in figure 4P. Auxiliary punches are

FIGURE 4P.
Removable tool holder

mounted in the die set with independent controls for fill, press and ejection positions. The compression stops for the secondary punches are slide blocks mounted on the stationary platen of the die set, and are moved out of the way at ejection. The correct fill for each section of the multi-level part is achieved by first raising the die for the proper fill ratio of the part section above the fixed or stationary punch, and then by independent adjustment of the secondary or auxiliary punches for their respective fill positions.

Figure 4Q illustrates a Class IV part configuration compacted by this method. During compaction the upper punch enters the die cavity and the die descends, escorting the secondary lower punches to their respective positive stops at final compaction. Powder is transferred into the secondary upper punch by proper timing of the die movement. There must be an upper core member, as in the multiple motion opposed ram system, to assure powder confinement during transfer. Ejection is by the continued downward motion of the die. The positive punch stops slide out, so that all levels of the part are fully supported until each level is free of the die cavity.

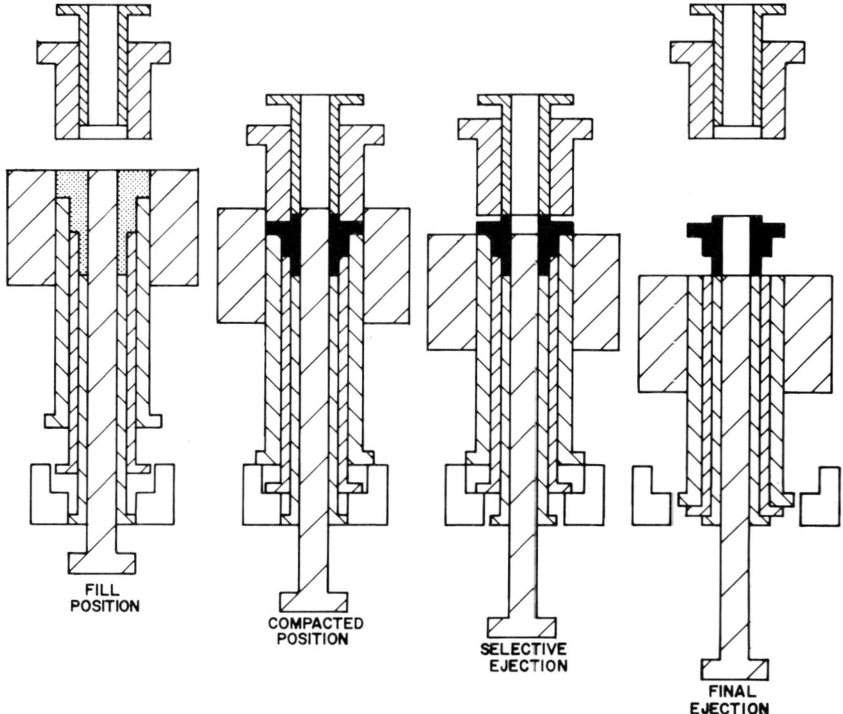

FIGURE 4Q.
Withdrawal press system multiple punches

4.7 ROTARY PRESSES:

A rotary press such as shown in figure 4R is a mechanically operated machine which uses a number of identical sets of tools to produce parts at high production rates. Some rotary presses can produce up to 10,000 parts per minute, but normal production rates for most P/M applications are up to 1,000 parts per minute.

The tool sets, normally called tool stations, are held in a head or a turret which rotates continuously. The sockets in the head which guide the upper punches, lower punches, and dies are precision bored so that tools with as little as 0.0005 in. (0.00127 cm) punch-to-die clearance can be used. The upper and lower punch sockets can be provided with hardened steel bushings which are replaceable. A hardened die table plate also can be used to reduce wear on the table surface. Figure 4S illustrates a typical rotary turret or head.

FIGURE 4R.
Rotary press

FIGURE 4S.
Rotary turret

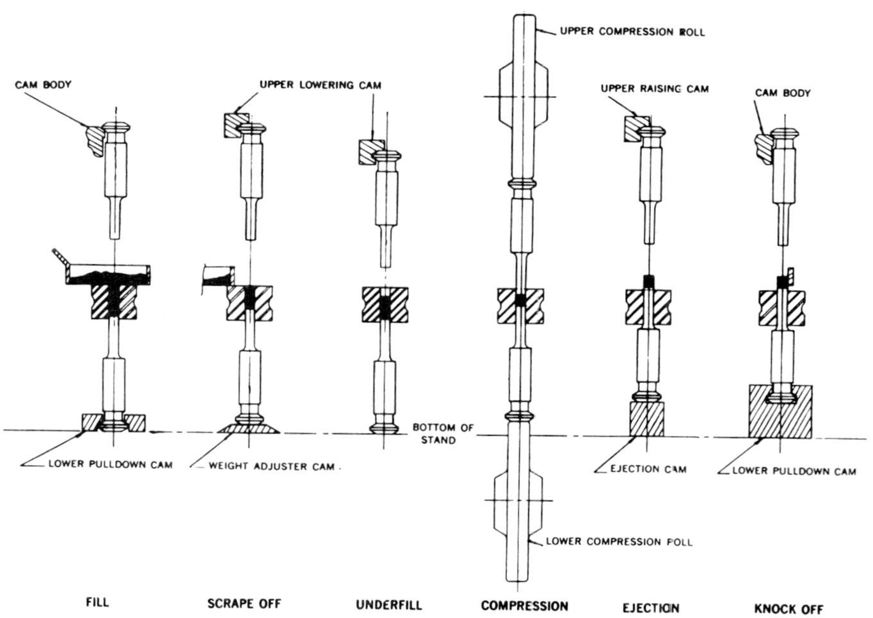

FIGURE 4T.
Rotary compacting system

The operation of a rotary press is relatively simple. The rotation of the head pulls the upper and lower punches past fixed surfaces called cams, and a set of pressure rolls that impart compression force from above and below. All rotary presses are double action. The design of the cam surfaces moves the punches up or down to provide the cycle of die filling, weight adjustment, compression, and ejection. This sequence is illustrated in fig. 4T.

Unlike a single punch press, the feed frame on a rotary press is held stationary. As the lower punches move under the feed frame, the punches are pulled down to form the die cavity into which the powder falls by gravity.

An inherent feature of a rotary press is "overfill-underfill". The lower punches are pulled down a fixed distance every time they go under the feed frame. This distance is usually more than the depth of fill required for the part. Consequently, the lower punches travel over an adjustable cam (weight adjuster cam) which causes the excess powder to be pushed out of the die cavity back into the feed frame. After leaving the weight adjuster cam, the lower punches drop down, resulting in a drop of the powder in the die cavities below the top surface of the dies.

Following is an example of this inherent overfill-underfill feature. If the required depth of fill to make a part is ½'' (1.27 cm), but the pull-down cam is designed to pull the lower punches down a distance of 1'' (2.54 cm), the lower punches are pulled down and the die cavities are filled with powder to a depth of 1'' (2.54 cm). . . overfill. The lower punches go over the weight adjuster cam, which is adjusted to cause the punches to travel up ½'' (1.27 cm) to a point below the top surface of the die. This action pushes powder out of the die cavity and back into the feed frame station.

After the lower punches leave the weight adjuster cam, they drop down to the initial fill reference of 1'' (2.54 cm). The ½'' (1.27 cm) of powder left in the die cavities also drops down. . . underfill.

The overfill-underfill feature and the constant movement of the powder in the feed frame usually gives much better weight control for a particular part than can be obtained on a single punch press.

Besides weight adjustment, another adjustment on a rotary press is for part thickness. This is done by adjusting the height of the lower pressure roll. By adjusting the roll up, the part becomes thinner. However, rotary presses can be furnished with an adjustable upper pressure roll system to control upper punch upper punch entrance into the dies. This is necessary if a flanged part is made with stepped or shouldered dies. Another feature of rotary presses is that all adjustments can be made while the press is in operation.

Rotary compacting presses can be furnished in tonnage ranges up to 100 tons. They can be modified to include stationary and movable core rod tooling. The movable core rod modification can be made for either a filling or ejection type core rod motion, or a combination of both.

5.0 HOT FORMING

P/M hot forming produces parts which are stronger than conventional cold pressed and sintered parts, and to tolerances and finishes that, in most cases, require little or no subsequent machining. This system usually includes a conventional compacting press which produces P/M preforms. The preforms are not necessarily even approximately to final shape, but must be of correct weight. Further, preform weight must be reproducible within narrow limits.

The preforms are then sintered in equipment as described in Chapter 11.0 of this manual, or in special equipment supplied as part of the hot forming system. Either before or after sintering, depending upon the equipment used, preforms are coated with a lubricant which aids in forming and helps to protect them from oxidation if exposed to the atmosphere.

After sintering, the preform is fed automatically through a heating unit which heats the preform to a controlled temperature between 1300° - 2030° F (704° - 1110°C), depending upon the material, part shape, and required final properties. The preform is transferred into the die area of a press and is formed in closed dies to its final shape with one stroke of the press. The formed part is transferred from the die to a conveyor or container, and on to final heat treatment and machining, if required.

Production speeds of five to thirty parts per minute can be produced by this process, depending upon part configuration.

5.1 PRESS TYPES:

Hot formed P/M parts can be made on hydraulic, screw, impact, toggle, or eccentric type presses. To insure minimum tool contact time, hence heat transfer from part to tool, precise control of the press positions and tonnage is essential. If clamping dies are used, hot P/M forming presses are usually of the mechanical eccentric type with auxiliary hydraulic motions for parts which require some tooling members to have a tonnage controlled dwell.

When slower production rates are dictated by part size, handling, or heating, it is customary to run the press intermittently, rather than at slow speed, in order to minimize tool contact times.

The motions required for a typical part are shown in the forming sequence in the forming press tooling section (fig. 5A). The motions shown in this forming sequence require a double action press, which has three upper and three lower tooling members. The upper die or clamping die is controlled hydraulically after it is brought into the clamping position by a mechanical motion of the forming press.

The hydraulic clamping is accomplished by mechanically closing a hydraulic valve and using the relative motion between platens to build a controlled clamping tonnage. The upper punch follows the main eccentric motion on the down stroke, but dwells on the up stroke while the part is stripped from the core rod and die. The core rod either has the same motion as the upper die, or the same motion as the main eccentric. The lower die and core rod, when used, are stationary. The lower punch has an adjustable press position and ejection position. If no upper outer die is required, a single action press can be used.

The single action press has the same lower motions as the double action, but only upper motion for the core rod and upper punch, with either mechanical or hydraulic stripping for the punch. Photographs showing typical die sets for single action and double action are shown in figures 5B and 5C.

Compacting Presses: The size, weight, and density distribution requirements of the preform are similar to those of conventional P/M structural parts. Size control is important for ease of transferring and proper material flow during forming. The density distribution is important in order to get proper heating

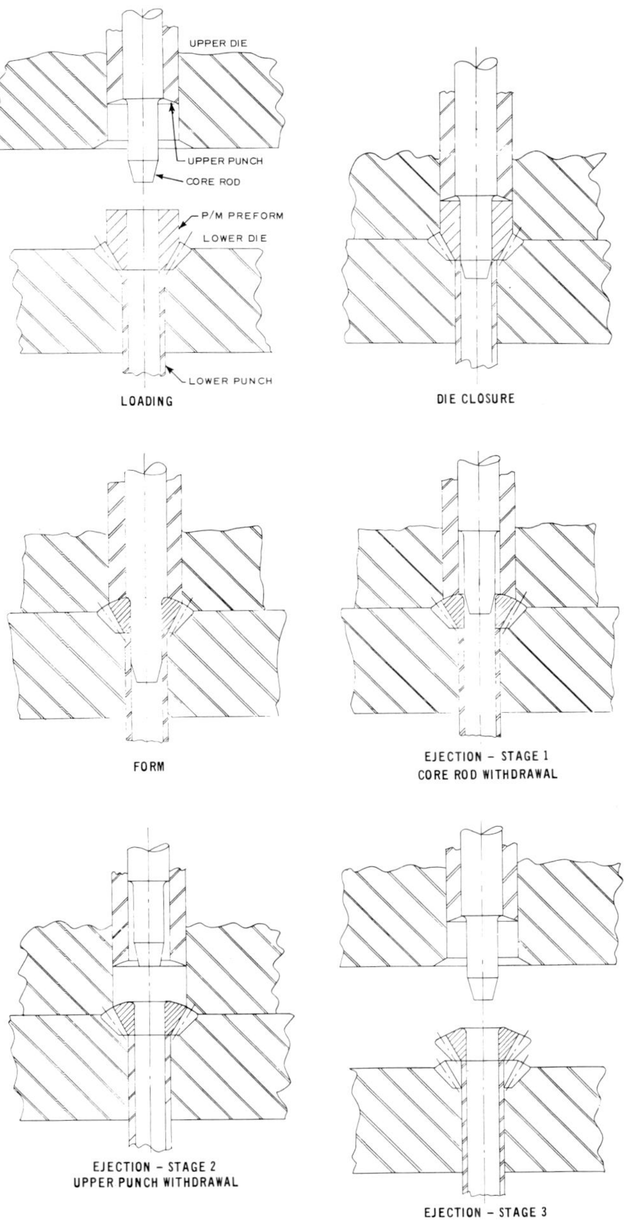

FIGURE 5A.
P/M hot forming - sequence of operation (cont.)

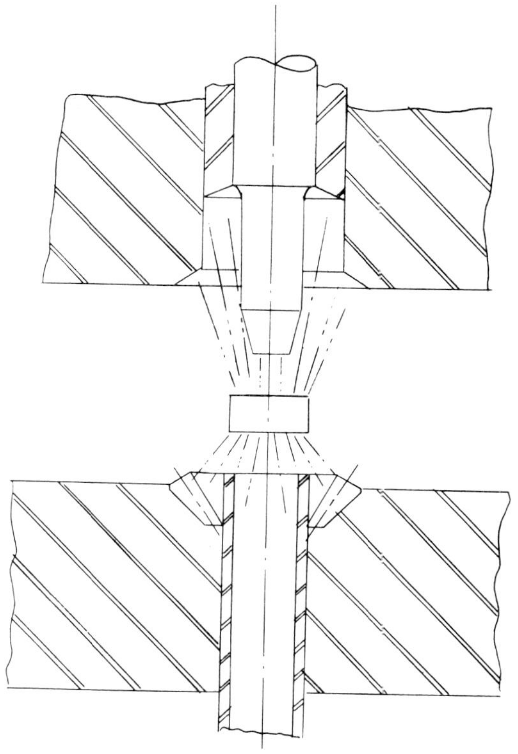

COOLING AND LUBE

FIGURE 5A.
Sequence of operation (cont.)

and material flow during forming.

The most critical factor in hot forming is weight control. Because the forming process occurs in a closed die, no excess material is squeezed out. Any variation in the preform weight is reflected directly in the weight and shape of the formed part. For example, a one gram change in weight will cause a 0.002 - 0.0035 in. (0.0051 - 0.0089 cm) change in height of a differential pinion gear of medium size. In order to provide adequate control over the weight variable, most presses making preforms for hot forming are equipped with electronic monitoring devices that reject preforms which are outside weight limits.

FIGURE 5B.
Die set
(Single action)

FIGURE 5C.
Die set
(Double action)

5.2 PART HEATING:

Prior to hot forming the preforms are brought to the correct temperature by induction heating. The preform temperature is monitored and the power level of the induction unit is adjusted to insure correct preform temperature.

The part shape, heating coil design, production rate, and frequency of the induction power are factors influencing the temperature and temperature distribution in the preform. Either solid state units or motor generators are used to provide induction power.

5.3 TRANSFER MECHANISMS AND FEEDERS:

Automatic handling of the preforms and parts is necessary because they are normally transferred at high temperature and at high production rates. The feeder mechanism feeds the part from a load station through the induction coils. The transfer mechanism transfers the preform from the induction heating coil to the hot forming die, and subsequently the part from the die to a receiving station. As a safety factor, these units are interlocked with the forming press either mechanically or electrically.

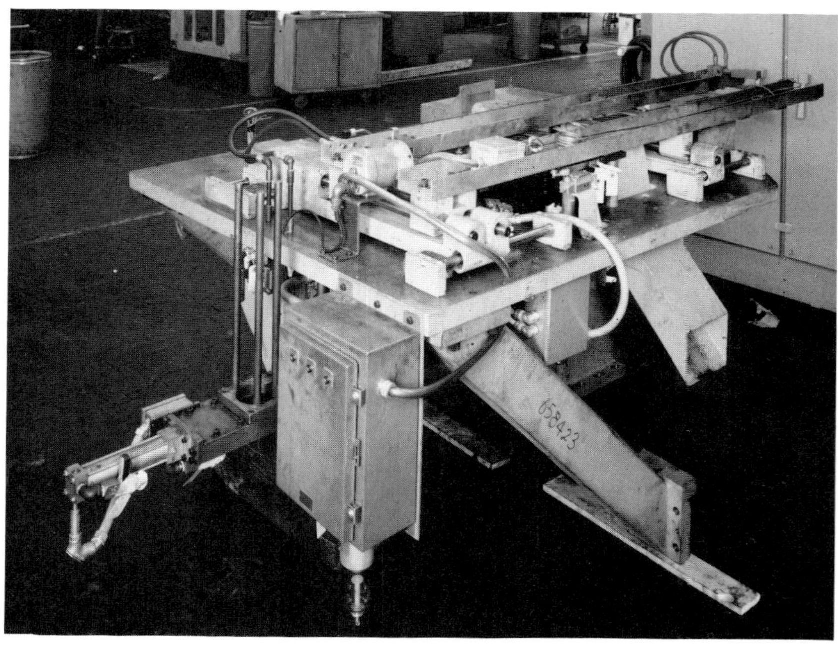

FIGURE 5D.
High speed transfer mechanism

Since each different part shape will probably require a different shape preform, changes must be made in the feeder, transfer mechanism, and possibly the induction coil each time a different part shape is to be formed.

The forming press transfer mechanism and feeder should be adaptable to accept different shape preforms and parts.

Two types of transfer mechanisms are shown. The first figure 5D is a high speed mechanism to be used with a continuously running forming press. It is mechanically connected directly to the main drive of the press, and its motions are controlled by a barrel cam. The second transfer mechanism, figure 5E has its own power source and is electrically interlocked with the forming press, allowing the press to run intermittently to slow the production rate while maintaining short tool contact times.

FIGURE 5E.

Transfer mechanism electrically interlocked with forming press

5.4 TOOLING, GENERAL:

Tool design for hot forming is complex. In addition to the consideration given to compacting press tools, there are the new variables of higher temperature and greater tonnages.

To minimize the heat transfer from the part to the dies and thermal stresses due to the sudden contact of a hot preform with cool tools, the tools are preheated. However, the tools are continually receiving heat from the preforms. To maintain a nearly uniform tool temperature, a liquid spray normally is used to cool the tooling members. After the dies are cooled a thin coating of lubricant is often applied. This acts as a thermal barrier to help slow the transfer of heat from the preform to the tools. It further aids in forming the part because of its lubricating properties.

Figure 5A shows, schematically, the tools and motions required to make a typical hot formed part. It also illustrates a die set press with position adjustments and controlling motion located on the press.

A removable die set is an important feature in hot forming because of the time required to allow the tools to cool off and warm up. When a tool change is required, a second die set with tools can be delivered to the area and preheated while the first tools are still producing parts. The die sets are usually constructed to allow for individual tool removal while in the machine. If simple tooling is used and one tooling member wears faster than the others, it is possible to cool the individual tool down with a water spray and change it without disturbing the other tooling members.

Dies: Because of the higher tonnages needed, shrink fits may have to be used in the female dies. Rules used for computing die shrink factors may not apply because of temperature variations within the die. Tool makers and hot forming system consultants can provide information on shrink ring data. In choosing a material for dies, consideration must be given to the fact that even though the dies may be preheated to an average temperature between 300° - 500° F (150° 260° C), the surface in contact with the preform may fluctuate between 300° - 1200° F (150° - 650° C). Several tool steel die materials which have been used are M-4, H-12, H-11, and H-17.

Punches: Punches should be kept as simple as possible. Thin protrusions to form chamfers and steps to raise short hubs or counter bores are possible. They must be backed up by other tooling members with the same motion during final forming. These shapes introduce lateral forces in the punches and make it difficult to remove the formed part from the punch. Also, the life of these tooling members will be shortened because of added temperature fluctuations and movement of metal along the surface. Punches can be made of the same materials as the dies.

Core Rods: Core rods have a large area exposed to the hot preform and little means of cooling. A hot formed part contracts rapidly after it is formed. It is easy to eject from the die, but difficult to strip from the core. The P/M compacting practice of waiting until the part is ejected from the die before stripping the core rod to lengthen core rod life is not applicable in hot forming. The core rod should remain in contact with the preform and part for the minimum time possible, and should be among the first of the tooling members to leave the part during ejection.

Tooling Clearance and Sizing: The clearances needed between tooling members depends upon desired concentricity levels, the configuration of the tools, and, to some extent, the production rate. An average value of 0.001 inches (0.0025 cm) per side at operating temperature would be a starting point.

In determining the size of the tooling members, note that the part shrinks after forming, while the die expands due to load and increased temperature. The core rod runs hotter than the other tools. The temperature for calculating part shrinkage is not the preform temperature as it goes into the die, but the temperature of the part as the forming press passes bottom dead-center. For example, to find the cold die. I.D. for a typical medium size part, the following computations were made from the cold finish part O.D.:

1. part shrinkage after forming	−0.0163	(−0.0414 cm)
2. die expansion due to load	+0.004	(+0.0101 cm)
3. die expansion due to temperature	+0.006	(+0.0152 cm)
total	−0.0063	(-0.0161 cm)

The cold die I.D. must be made 0.0063 in. (0.0161 cm) larger than the cold finish part O.D.

5.5 HOT SIZING:

Hot sizing is very similar to the P/M forming process and the previous discussion on hot forming equipment and tooling applied to hot sizing. The two processes differ by how closely the preform resembles the finished part.

The hot sizing operation requires a preform of final shape and dimensionally close to the required part. There is a minimum amount of metal flow. Tolerances and finishes are improved, but physical properties are not as good as those obtained by hot forming. The press tonnage required is considerably less than required for hot forming, and tool life is greater. For parts which require physical and/or mechanical properties intermediate between hot formed and hot sized parts, the preform shape can be modified. Such preforms are designed to give only the amount and distribution of strength required. This technique minimizes the amount of press tonnage and die wear for medium strength parts.

6.0 CONVENTIONAL COINING AND SIZING

Some parts are repressed after sintering, or partial sintering, for better density and strength to improve surface finish, and to establish closer dimensional tolerances. Restriking to improve density or surface condition is usually termed coining, and restriking to improve tolerances is normally termed sizing.

Parts can be repressed on compacting presses, modified presses, or presses specifically designed for repressing.

In general, coining presses have higher production rates than compacting presses for the same part size. The limitations on production rates are feeding and removing parts from the die and the allowable temperature build-up in the die due to friction.

The coining or sizing press feeding mechanism must be capable of high speeds as well as extremely accurate operation. A popular restrike feeder is the indexing dial type. This feeder is usually mechanically driven by the press motion and brought into final position with a shot pin or a precision cam box. A restrike press with a high speed dial feeder is shown in figure 6A.

It is important that checks be incorporated into the automatic feeder to insure that there is one part and only one part in the die for each stroke of the press.

FIGURE 6A.

Restrike press with high speed dial feeder

FIGURE 6B.
Coining press

Another type of coining press is illustrated in fig. 6B. This machine utilizes the removable die set concept of tooling and can be used for coining, sizing, or compacting Class I and Class II parts. The transfer mechanism which is connected and driven mechanically by the press mechanism is a four-station walking beam linear system.

Unlike hot forming, tool contact times are not as critical in restriking and variable speed presses are advantageous to obtain maximum production possible for the part size and shape.

7.0 COMPACTING PRESS GENERAL INFORMATION

7.1 POWDER FEEDING:

Automatic powder feeding and part removal are two prime requisites of any P/M compacting press. Both of these functions are normally achieved by the press powder feeding system. The same feeder which supplies powder to the die cavity is also used to remove the compacted part from the point of operation area on the press die table.

There are three basic types of feeders used on compacting presses. One type, normally used on moving die table press systems, provides a straight in-line reciprocating action over the die with the feed shoe connected directly to the supply hopper which pivots on its mounting. The feed shoe motion is imparted by a direct connection of the shoe with a mechanical, pneumatic, or hydraulic force system (fig. 7A). Additional agitation of the shoe can be provided by contouring the feeder cam in a mechanical system or by pulsating the penumatics or hydraulics when using fluid power.

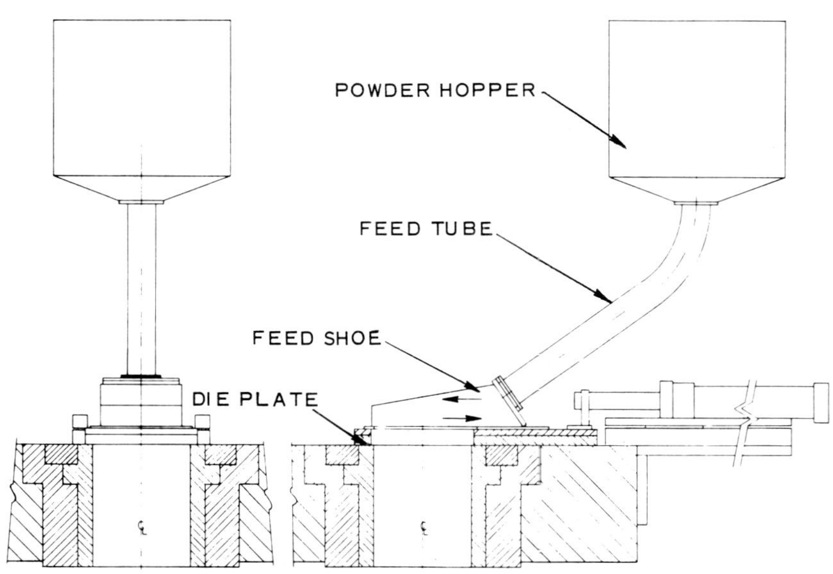

FIGURE 7A.
In-line direct-fill shuttle feeding system

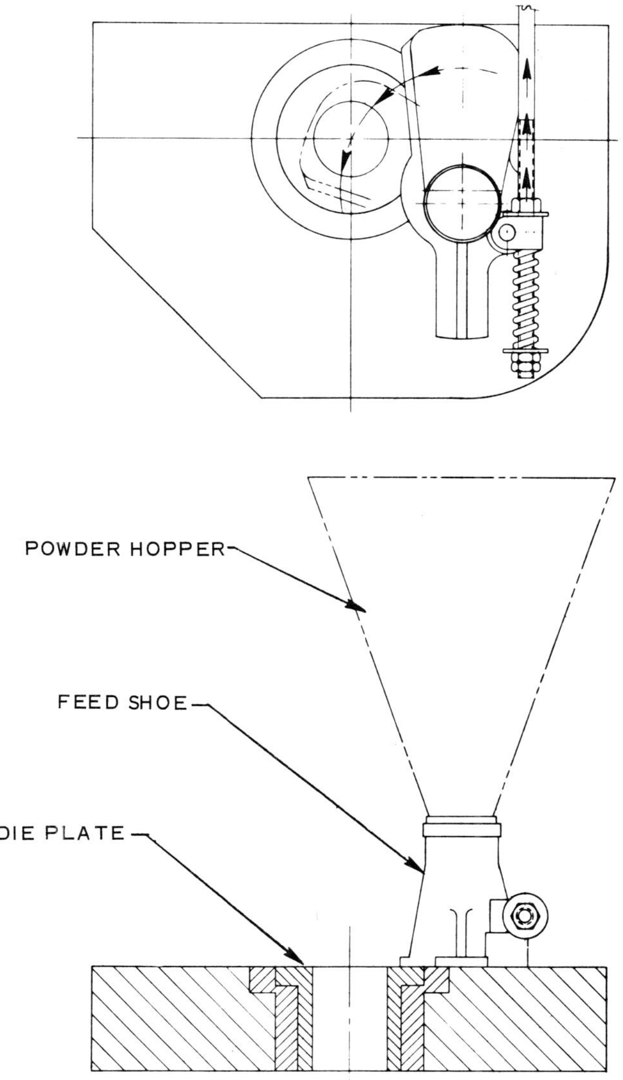

FIGURE 7B.
Arc type powder feeding system

Another type of feed system which is normally used on mechanical presses with a stationary die table uses a pivoting action of the feed shoe over the die area. This system is commonly called the arc type feeder. The shoe is pinned or guided on the die table while its movement is controlled mechanically with the base press motion (fig. 7B). The hopper is mounted directly above the feed shoe, providing a constant pressure head of feed material into the shoe. Agitation of the shoe is usually accomplished by contouring the press feeder cam.

The third type of system commonly is called a shuttle feeder. Its motion is the same as the in-line system. However is does not have a direct connection with the supply hopper. At fill it moves with a metered amount of material from a position under the hopper to a position over the die cavity. At the same time it closes off the powder coming from the supply hopper with a cut-off plate. Its motion is imparted by a direct connection of the shoe with a mechanical, pneumatic, or hydraulic force system (fig. 7C). This system supplies the same metered amount of material to the die cavity on each press stroke.

In general, it is desirable that the feeder cover the die opening while the bottom punch is at the eject position. This sets up a condition to minimize air entrapment, which could hinder the powder feeding action. If the bottom punch has a projection on it, it is impossible to completely cover the die with the feed shoe while the lower punch or die are at ejection. However, it is still desirable to set up the feeder timing to allow for filling as soon as clearance

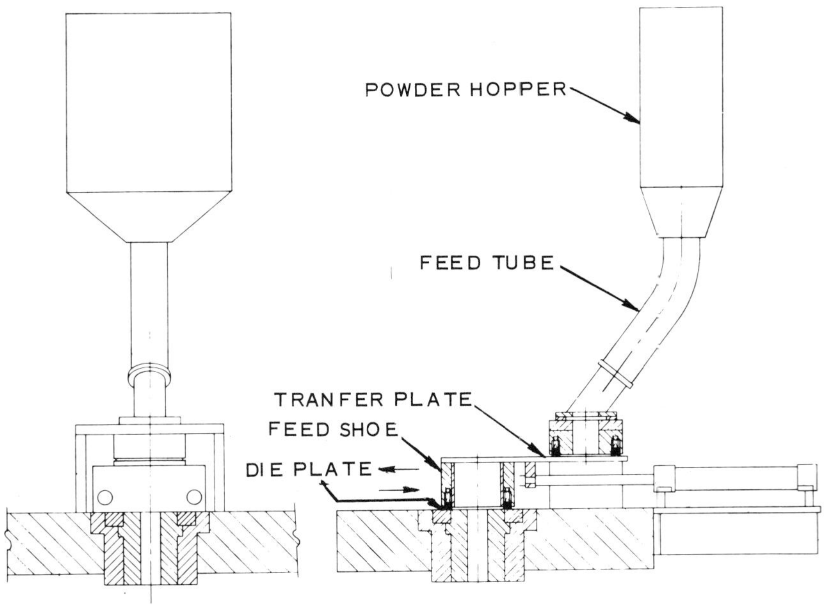

FIGURE 7C.

In-line shuttle powder feeding system

can be achieved over the lower punch, preferably before the tool system has reached its fill stop levels. Most press manufacturers supply auxiliary feed shoe agitation attachments to aid filling of material under difficult conditions.

It is important to utilize electrical interlocks on all feed systems using pneumatics or hydraulics to drive the feed shoe. These prevent inadvertent system failures which might occur through loss of synchronization in timing of the feed shoe to the press ram.

7.2 COMPACTING PRESS ATTACHMENTS:

Most press manufacturers offer with their standard compacting equipment, a number of press attachments which serve to aid the press or tool system in the performance of its task. These attachments can be subdivided into two major categories:

(1) Auxiliaries which serve to complement the basic pressing machine.
(2) Auxiliaries which serve to aid the press and the tooling system in conjunction with a specific part application.

General Systems: Automatic lubrication systems, air clutches, hydraulic equalizers, specialized electrical control systems, and carbide wear parts in critical areas all serve to complement the basic pressing machine.

Selection of the automatic lubricating system (refer to press maintenance section) will depend on the user's in-house maintenance procedures. The air clutch, when offered on a mechanical press as an option over a mechanical clutch, provides additional safety and reliability when there are continuous starting and stopping operations as encountered in coining and sizing.

The installation of carbide buttons or wear plates in the gib-crosshead area on either mechanical or hydraulic presses will prolong the working life and alignment. Their selection and use depends primarily on in-house operating conditions.

Automatic conveyor belts or part take-off devices, when offered for part removal from the press, should be considered seriously in terms of added safety and efficiency of the press operation. Motorized upper ram, raising and lowering systems on large tonnage presses should be considered as a definite aid in reducing tool set-up times.

Auxiliary attachments which serve to aid the press and the tooling system in conjunction with a specific part application are:

Upper Punch Hold Down: This attachment normally is offered by all press suppliers. It usually consists of a pneumatic, hydraulic, or mechanical device built into the upper press ram or one of the main compression members, to provide added top pressure on the part during the ejection portion of the pressing cycle. Added pressure at this point in the press cycle effectively increases the top punch dwell on the part, it allows more time for entrapped air in the green part to escape. When compacting materials such as ferrites and carbides the hold down system is essential, as entrapped air will produce part laminations.

Hydraulic or Pneumatic Equalizer Systems: This attachment usually is offered on mechanical presses to insure compacts of even density despite variations caused by irregularities in fill of material being compacted. It normally consists of a hydraulic, mechanical or pneumatic link in one of the main compression members of the press. The system is pre-set to the desired operating pressure for the compact. If this pressure is exceeded, deflection of the system link occurs, preventing the application of further pressure to the part.

The system can also be used as an overload release mechanism where the setpoint for the system link is established at a point higher than the necessary pressure for the part. This provides a margin of safety when overpressing. Uncontrolled material fill or broken green part fragments could cause damage to delicate tool members.

For this application, the release mechanism is usually linked electrically with the press clutch mechanism. As the setpoint is reached, the clutch is de-energized and the brake is applied, stopping press motion.

Movable Core Rods: This option is usually offered on both hydraulic and mechanical presses, as either type can be equipped with an independent core rod action. Motion of the core rod can be from mechanical, hydraulic, or pneumatic energy sources. The movable core system is normally utilized to reduce ejection loads, to aid in filling, or to provide the relative motion necessary for compaction of complicated shapes. It also helps to hold dimensional tolerances of parts with long, small diameter cored holes.

The core rod mechanism is normally mounted on the lower portion of the press frame. The simpler types usually provide a standard two position adjustable stop motion in one of three operations:

A. Filling and Stripping: In this operation the core rod descends before the press reaches the fill position in order to take on added material into the die cavity. It is then raised prior to the completion of the fill motion so the excess material can be pushed back into the feeder shoe.

During compression the core rod remains neutral and at ejection it moves toward its lower stop. This action results in stripping the core rod from the piece, and places the core rod in the initial overfill position.

On presses which utilize the lower punch for ejection motion, this results in the added advantage of a difference in relative motion at ejection. The core rod is moving down while the lower punch is moving up, which serve to decrease some of the ejection load on the press. The system can also transfer powder into top hub part shapes when dual upper punches are utilized. In this case the transfer time is delayed until the upper punch starts its initial entry into the die.

B. Ejecting: With the ejecting type movable core rod, the core rod is held in position at its lower stop level during fill and compression. At ejection, it is timed to raise to its upper stop limit with the compact above the die table. The part expands off the core rod and the rod is retracted to its lower limit. This action also results in the decrease of ejection loading on the press and wear on the core rod.

C. Floating Core Rod: The movable core during this operation is pre-set to a certain fill position and is supported in this position by mechanical, pneumatic, or hydraulic pressure. During the compression portion of the press cycle, the core rod floats back to a positive mechanical stop as the piece is being compacted. At ejection the core rod is returned to its original fill position.

The floating core system, when applied to a stepped core rod configuration, provides relative motion between the lower punch and core rod, to help eliminate shear planes and crack formations in the part. On pneumatic and hydraulic systems, the floating core has the added versatility of variable resistance during compaction through the control of fluid pressures. When applied to small-diameter long-length core rods, the system helps to reduce distortion of the hole in the finished part. This is achieved by reducing the compressive stress on the core rod members, which if excessive on high l/d components could cause bending during compaction through the floating action of the system.

On presses 100 tons and over, more sophisticated three and four position core rod systems are offered. Basically they provide a combination of the motions previously described. The four position core rod has two separate stop systems. In one case the core rod could be utilized as a filling member for the fabrication of top part hubs, with its adjustable stop locked in position for compression. After compression the stop is unlocked and the core rod member serves as a stripping tool to aid in ejection of the part.

Overfill-Underfill: This option is generally available on either hydraulic or mechanical presses. On overfill, the press is provided with an adjustable stop mechanism to take on more powder at the initial portion of the fill section of the cycle. Prior to completion of fill, while the feeder is still over the die cavity, the overfill mechanism is released so the die or lower punch can return to the correct part fill position. This addition and removal of extra material during fill results in improved powder-fill characteristics for hard-to-fill shapes such as high l/d ratio parts and fine tooth gears.

The underfill action is the direct opposite of the above system. In this operation the powder charge, after filling has taken place, is usually dropped to a level lower in the die by moving an adjustable stop member controlling the die table or lower punch. This allows the upper punch to seal off the die during the initial compression phase of the press cycle without displacing powder.

Dual Upper Punch Systems: This option, which can be mechanical, hydraulic, or pneumatic, is used when the part to be compacted requires the forming of an upper hub on a flange type part. An inner upper punch and an outer upper punch are used to form the top of the hub and the top of the flange. Most systems have adjustment for fill compensation on the outer punch, control of resistance to float-back of the outer upper punch during compaction, and an ejection motion behind the inner upper punch to eject the hub out of the outer upper punch.

7.3 LUBRICATION:

Lubrication systems can be classified as manual or automatic.

Manual lubrication systems can be individual point or centralized point systems. The individual point system requires servicing by a handgun at each point of lubrication. The centralized point system employs a pump to force-feed the lubricating medium, under pressure, to the individual points throughout the press. The pump is manually operated and may or may not be attached to the press. Individual point systems, almost exclusively use grease as the lubricating medium. The centralized point system can be set up to operate with oil or grease.

Automatic lubrication systems are centralized point systems equipped with a pump which can be programmed to automatically force-feed the lubricant to the individual points throughout the press. The pump can be driven by an electric motor, pneumatically operated or driven from some moving member of the press.

Information on the frequency of lubrication and the type of lubricant can be obtained from the machine manufacturer.

7.4 PRESS MAINTENANCE:

Preventative Maintenance: A preventative maintenance program can, if performed on a regular basis, pay for itself by uncovering minor problems which, if not found and corrected, could lead to major repair problems.

The type and complexity of the press will determine the format of the preventative maintenance program to be followed. The instruction manual supplied with the machine may outline the program; if not, the machine builder will aid in establishing one.

A preventative maintenance program for mechanical presses should include provisions for a daily, weekly, and monthly inspection and servicing of the lubrication points, inspection of cams and cam rollers, and a visual inspection of bolts and screws. The monthly inspection should include, in addition, a general cleaning of the press.

A program for hydraulic presses should include a daily inspection of the hydraulic oil level, temperature, and pressures. Provisions should be made for daily, weekly and monthly inspection and servicing of the lubrication points.

The weekly inspection should include cleaning of the hydraulic oil strainer, filters, and magnetic traps. The monthly inspection should also include a general cleaning of the press.

7.5 SAFETY:

Safety is the joint responsibility of the machine manufacturer, the user, the operator, and set-up and maintenance personnel.

The manufacturer's responsibility is to provide a machine which, when operated and maintained properly and used for its intended purposes, will not subject the machine operator to risk of injury. Instructions for normal safe operation and recommended maintenance are also part of the manufacturer's responsibility.

The user's responsibility is to insure that the operator, set-up, and maintenance personnel follow the operating and maintenance instructions given by the machine manufacturer. The user should be sure that all personnel observe and obey any warning signs or labels that the manufacturer installs on his equipment. The user must not alter, or allow his personnel to alter, the machine cycle, electrical, hydraulic, or pneumatic circuits without first obtaining clearance from the machine manufacturer.

The user also should evaluate each of the operations to be performed on the press in order to determine the proper point of operation safeguarding. He should then make available this proper point of operation safeguarding and insure its use.

For more detailed information, a standard entitled "Safety Requirements for the Construction, Safeguarding, Care and Use of P/M Presses" can be purchased from the Metal Powder Industries Federation, P. O. Box 2054, Princeton, New Jersey 08540. The standard applies to mechanical and hydraulic compacting presses which are designed, modified, or converted to compress metallic and non-metallic powders.

PART
TWO

TOOLING

8.0 TOOLING AND TOOL DESIGN FOR POWDER METALLURGY

Every part being considered for powder metallurgy must be analyzed thoroughly. First, the material specifications and pressure requirements must be determined in order to fit the capabilities of the equipment. Then, the motions necessary to produce the part must be determined in order to utilize most effectively those furnished by the press and those that must be incorporated in the tooling.

Good tooling design is a key element in a part well made at a minimum of expense. The high investment in compacting equipment requires an appreciation of this necessity coupled with the benefits of experience.

The following section on tooling and tool design for powder metallurgy contains basic technical data to assist the engineer in designing sound tooling within his equipment and to outline for the purchaser of powder metallurgy and other parts made from compacted powder the design parameters of the process so that he may better understand the capabilities and limitations of the custom parts manufacturer.

8.1 THE DIE:

Most presses have large die openings. The main reasons are to provide ample space for adapting tools from one make or size of press to another, and to allow room for shrink rings when insert type dies are to be used.

In checking die wall thickness, it is assumed that full hydraulic transmission of pressure is obtained, even though this is contrary to the no-side-flow theory. It is also assumed that the tensile stress in the die wall is distributed over an area corresponding to three times the thickness of the piece.

Die inserts may be held in place by clamping or by shrink fits. The cases for shrink fits are preferably made of steels, which are not as hard as the usual die steels but are much tougher such as: chrome nickel tool steel. This steel is also used when the die insert is to be held in by screws as the toughness minimizes thread failures. When making shrink fits the usual interference between the

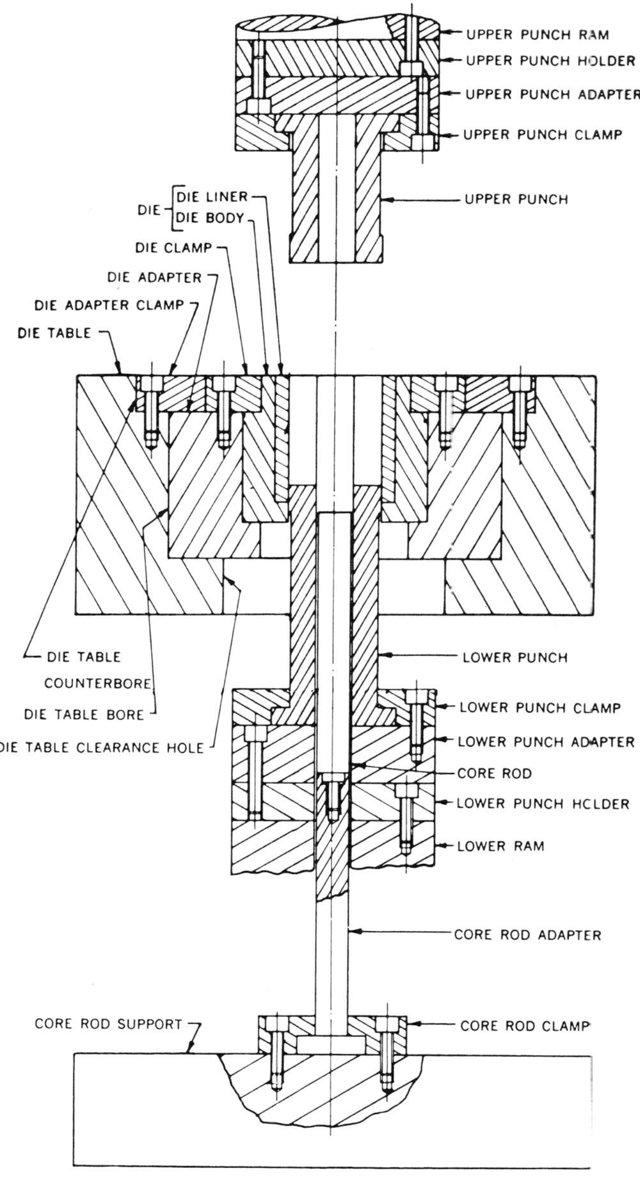

FIGURE 8A.
Cross section view of tooling for powder metallurgy compacting

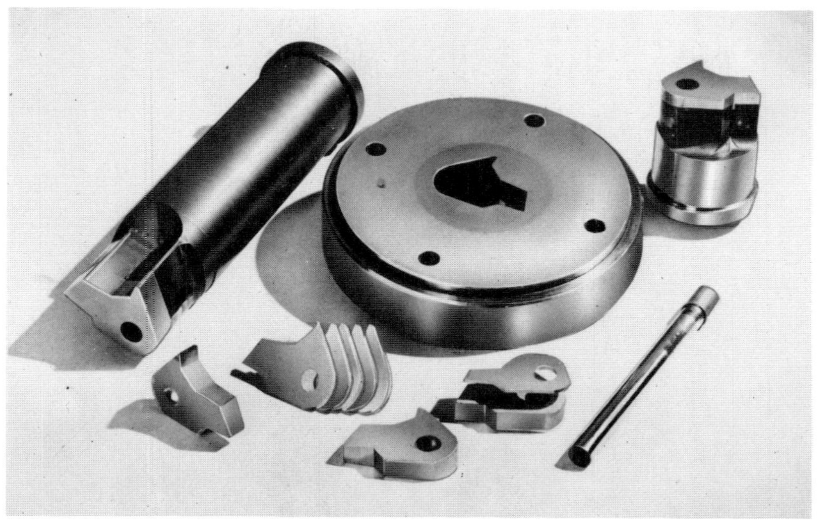

FIGURE 8B.
Typical tool set for making the part shown in lower left

steel or carbide inserts is 0.002 inch per inch of diameter. Use as large a case diameter as possible to support the insert and prevent it from "breathing" and eventually failing. When carbide inserts are to be shrink fitted only about 0.0010 inch per inch should be allowed as the carbide will not "give" as much as the steel, carbides having only about one-third the elastic modulus of steel.

Die entrance edges should be beveled (about 15° from the vertical) or radiused. The bevel should be 1/32 inch deep, or less, depending on the size and the thickness of the pressed pieces. This minimizes injury to the punch faces when setting up and operating. When possible, horizontal joints in or near the pressing area of the die wall should be avoided, since fine powder works into such joints and spreads the sections vertically in spite of all precautions regarding finish of mating surfaces and high retaining pressures.

It is sometimes necessary to taper dies to aid in relieving expansion strains during ejection, otherwise horizontal laminations may appear in the compressed pieces.

When taper is required it is usually not necessary to make full allowance for the complete expansion of the piece—an allowance of 2/3 of the expansion usually being sufficient. If the pressed piece after ejection is 0.006 inch larger than the die at the compression point, the taper can be made 0.004 inch.

The useful life of the die depends on many factors; such as, the nature of the material being pressed, the unit pressure to be used, allowable tolerances in the finished compacts, material of construction and surface finish in the cavity. High chrome, high carbon steels are used for medium production re-

CEMENTED CARBIDES FOR POWDER METALLURGY TOOLING PROPERTIES AND TYPICAL APPLICATIONS

NO.	%. OF BINDER	HARDNESS, RA	TRANSVERSE RUPTURE, PSI	COMPRESSIVE STRENGTH, PSI		TOOL APPLICATIONS		
						CORES	PUNCHES	DIES
C-4	3%	92.3	177,000	800,000	↑			Bearing Dies
C-9	6%	91.5	230,000	710,000	SHOCK RESISTANT	Simple Shapes Short Lengths		Straight Thru Dies—Simple Cavity Contour
C-10	6 to 9%	90.6	280,000	650,000			Ceramics-Ferrites High Polish & No Face Projections	
C-11	12 to 13%	89.7	310,000	600,000		Step Cores & Complex Contours	Ceramics-Ferrites Metal Powders Simple Face Projections	
C-12	14 to 15%	88.5	340,000	580,000	WEAR RESISTANT	Step Cores & Vulnerable Contour		Complex Shapes Gear Forms Sectioned Dies
C-13	15 to 20%	87.4	375,000	550,000	↓	All Cores within Physical Limits of Carbide	All within Physical Limits of Carbide	Multi-Level Dies Vulnerable Projections
C-14	20 to 30%	82 to 86	365,000	470,000				

All property data represents average for grade

FIGURE 8C.

quirements. The analysis usually runs about 12% chromium and 2% carbon, and both oil hardening and air hardening grades are available. The air hardening type is used for dies having sharp cornered cavities which might not stand the shock of oil quenching. These dies should be heat treated to obtain a hardness of 60-64 Rockwell C.

When production volume is high or abrasive conditions are encountered, dies should be made of tungsten carbide with:

Low Cobalt 3-6%	Medium Cobalt 9-12%	High Cobalt 15-20%
round dies	simple contour	gear dies and difficult contours

When unusual shapes are encountered a number of carbide inserts can be fitted together with rings as shown in Figure 8D. This minimizes costly machining operations of solid dies.

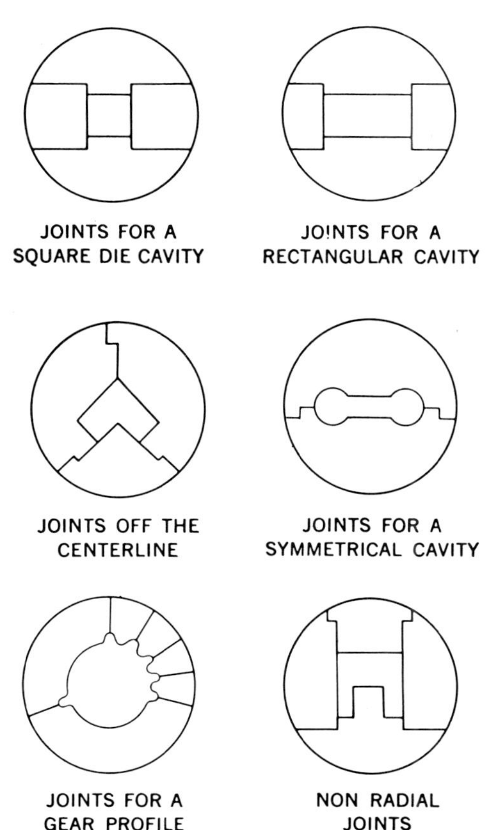

FIGURE 8D.
Carbide inserts

8.2 THE PUNCH:

Punch steel requirements are very different from die steel requirements. Here toughness is an important factor. High carbon, high chrome steels are too brittle in most cases; 3% nickel, 75% chrome and carbon around 0.40 to 0.50%— the lower carbon used when sections are thin or chamfered edges are present. Special analysis in the A.I.S.I. 3400 class meet the above specifications. For less delicate parts, the 320 class steel containing 1.75% nickel may be used.

When abrasion of punch face is high, punch inserts in chrome nickel steel holders can be used. The inserts can be made of high chrome steel with 1.5% carbon instead of 2% for simple shapes. If bevels and other stress concentrating details are present, the 5% chrome steels can be used. In those cases of multiple punch set-ups, where a punch may have a function partly as a die, the use of the 1.5% carbon high chrome steel or the 5% chrome steel is advisable.

Carbide punches are sometimes used. The grade used should not be as brittle or as hard as that used for dies. The 9% or 12% cobalt grades are more durable. When using carbide tips the back up steel should be low chrome rather than high chrome tool steel to minimize mushrooming of punches under pressure after brazing.

Punches and core rods are relieved 0.005 to 0.010 (0.12 to 0.24 mm) on the diameter and 0.0025 to 0.005 inches (0.06 to 0.12 mm) all around on all profiles to permit the escape of powder passing down beyond the punch faces. The actual close-fitting portions are made as short as is possible. Care must be taken, in considering the length of each fitting portion, to allow for relative motion between parts. Punches, particularly if they form chamfers, tend to chip at the edges, and may require regrinding several times in their useful lives. Some allowance must be made for this on the assumption that each regrind removes at least the length of the chamfer.

Core rods are used to form blind or shoulder holes. Consideration must be given to column loading and other tougher steels must be used in this application. The working length of a core rod should be held to a minimum.

Carbide coating applied by "flame plating" or reverse electrical discharge are often used on core rods for highly abrasive application.

8.3 TOLERANCES AND CLEARANCES:

Punches should be made to fit the dies within a specified clearance and not made to tolerances. The tolerances should be on the die and core rod dimensions (and the mating parts fitted with suitable clearances). The finest fits are required for making bearings and bushings. Any slight variation in powder fill tends to push the core rod to one side taking up all the clearances in one direction and producing eccentric bushings. No amount of subsequent pressure in the sizing operation can entirely correct the original eccentricity. For bushings the diametral clearances are usually not over 0.0002 inch. Eccentricity of I.D. and O.D. of punches should not be over 0.0002 inch T.I.R. For other applications the clearances are generally more liberal 0.0005 to 0.001 inch on diametral dimensions.

In the making of the punches themselves, concentricity of the punches and the punch shank or punch holder need not be held to high accuracy. Most presses include provisions for locating the punches concentric with the die so that alignment can be obtained. However, there is no provision for adjusting out-of-square and the squareness must be accurately maintained in the tools especially in close fitting dies for long pieces.

8.4 FINISHES:

Die cavities and core rod should be lapped or polished to a high finish after final grinding and the last polishing or lapping should be parallel to the axis of the tools. The microinch finish should be 5 or better. When surface finish can not be readily checked with a profilometer, visual check for ''mirror'' finish by an experienced tool maker is satisfactory. A well polished surface should have the same characteristics as a glass surface.

Punch faces and punch ''lands'' should have the same surface finish as the die cavities, and the final polish on the punch lands should be parallel to the punch axis. Punches and dies with poor surface finish wear out of tolerances much faster than when properly finished, and may prevent the tooling elements from moving freely to their proper position.

8.5 PUNCH AND DIE ADAPTORS:

With proper preliminary planning and ingenuity in adaptor design, large savings in tool cost and emergency scheduling can be made. By the use of proper adaptors the basic tool element can be held to a minimum size. Where a variety of presses are in operation it is advisable to provide punch and die adaptors so that tools from one press may be operated in other presses of different size or make.

The adaptors should be made from steels having adequate structural and dimensional stability so as not to detract from the accuracy of basic tool elements.

8.6 TOOLS FOR COINING OR SIZING:

Pressed and sintered compacts coming from the sintering furnace may be off size either intentionally or unintentionally. Such parts can be repressed, sized or coined to increase the density to reshape or to correct dimensional variations.

Repressing can occasionally be done in the same die or slightly different die in the same press, equipped with a part feeder instead of a powder feeder, or, for short runs the parts can be fed by hand.

In many cases the tools are very similar to the forming tools and are made of similar or slightly harder materials. Sizing pressure may run 50% to 100% greater than forming pressures. The die and core rod are provided with a tapering lead-in to assist the entrance of the parts into the die. In some cases the core pin is an integral part of the upper punch, or it may be a separate upper punch, or the hollow punch may be spring mounted on the core rod pin.

Another method, used chiefly in self-lubricating bearings, is to force the bearing into the die and then run a spherical burnishing tool through to size and refinish the inside diameter. Self-aligning bearings and other spherical parts require special treatment to remove the central flat left by the forming process. The spherical section of these parts is sized half in the upper punch, which comes down and meets the die but does not enter it. The upper punch has a thick wall to withstand the sizing pressures.

PART
THREE

ISOSTATIC COMPACTING

9.0 ISOSTATIC PRESSING

INTRODUCTION: Isostatic pressing is a P/M forming process that applies equal pressure in all directions on a metal powder compact, thus achieving maximum uniformity of grain structure and density. Simple or complex shapes requiring uniform strength and fracture resistance are the primary applications for the process. However, the process also lends itself to applications where larger P/M parts and more extreme ratios of length to diameter are necessary.

Isostatic compaction should be considered as an alternative to die-compaction. While die-compaction has remained the dominant method of pressing metal powders because of the economy with which complex shapes of accurate dimensions can be formed, it has some technical limitations such as size or configuration that restrict its use. The alternative processes, of which the isostatic techniques form one important group, are usually adopted only when die-compaction is ruled out.

Apart from advantages in forming powders that are inherently difficult to compact, isostatic compaction is often preferred for parts with large ratios of height to cross-sectional area. There are applications in which cold isostatic compaction is preferred on economic grounds; notably, when producing a small number of components where the high first cost of die-compaction tools cannot be justified, and some machining is permissible, or when very large compacts are needed. The flexibility of equipment for cold isostatic compaction is another advantage; a single piece of equipment can handle one large compact or many smaller ones. Compacts of different sizes and materials may be processed together, provided that the compaction pressure is acceptable for the shapes involved. Pressure can also be easily varied from one run to the next. A variety of materials can be pressed isostatically on a commercial scale, including metal powders, ceramics, plastics and composites. Pressures required for compacting range from less than 5,000 psi to 100,000 psi and above.

Sizes of isostatically pressed parts are limited only by the size of the pressure chambers, which to date have been made as large as 60 inches I.D. X 150 inches I.L. Configuration of the part is limited only by the complexity that can be designed into the flexible tooling. In addition, the process lends itself to production of thin-wall sections such as in tubes and crucibles. In many applications a minimum of finish machining is required.

9.1 ISOSTATIC PRESSING EQUIPMENT:

There are two types of isostatic pressing: the Free Mold or "wet bag" method and the Fixed Mold or "dry bag" method. They are identical in principle, achieve similar metallurgical properties, and require the same equipment. They differ, however, in the configuration and placement of the tooling.

Both methods utilize a pressure vessel with a closure device designed to contain a fluid under high pressures, along with a pressurization system and controls. Today's more advanced production-scale isostatic presses are equipped with various automated systems for powder fill, pressure vessel operation and part ejection.

The Free Mold Process is used in batch-scale production, in applications requiring complex shapes, and for most research and prototype studies. This technique provides flexibility in pressing a variety of sizes and shapes, limited only by the inside dimensions of the vessel. In Free Mold Isostatic Pressing, depicted in Figure 9A, flexible molds are first filled with the powder, and then vibrated, jolted or subjected to vacuum to remove entrapped air and assure optimum distribution of the loose powder. After filling, the molds are sealed water-tight and placed into an immersion basket or rack assembly for introduction into the pressure chamber. As many molds as will fit can be placed into the pressure chamber during a given cycle. The chamber is then filled with the hydrostatic medium (water or oil), the cover is closed and secured, and the vessel is pressurized to the required pressure.

The Fixed Mold Process, depicted in Figure 9B is characterized by the flexible mold positioned directly inside the pressure chamber. The mold may be supported by a perforated outer shell and with a lip on the mold sealing against the side wall of the chamber. An annulus is provided between the vessel wall and the outside surface of the mold to allow the hydrostatic fluid to enter during the pressing cycle. As with Free Mold Pressing, equal and simultaneous pressure can be applied to the mold.

The powder fill is made directly into the cavity of the fixed mold, which acts as an integral part of the pressure vessel. Hydrostatic fluid does not come into contact with the operator's hands. Evacuation of air prior to pressing if necessary, is generally achieved with a vibration device or vacuum connection in the pressure vessel cover. The sequence then proceeds as in Free Mold Pressing.

In Fixed Mold Pressing, one part is ordinarily pressed per cycle per pressure chamber. The process lends itself to such shapes as bars, rods or tubes.

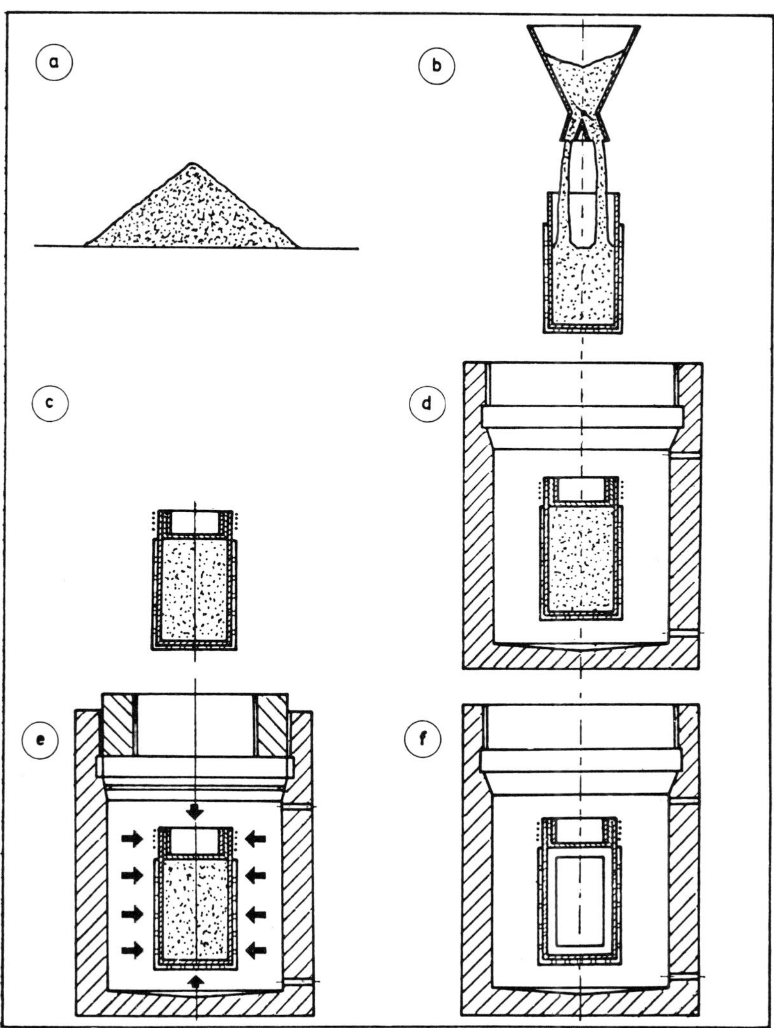

a) material to be compacted (powder)
b) fill-up of flexible form
c) closed and sealed form
d) form in pressure medium in pressure vessel
e) pressurising
f) resulting compact, after decompression

FIGURE 9A.
Schematic of isostatic pressing principle

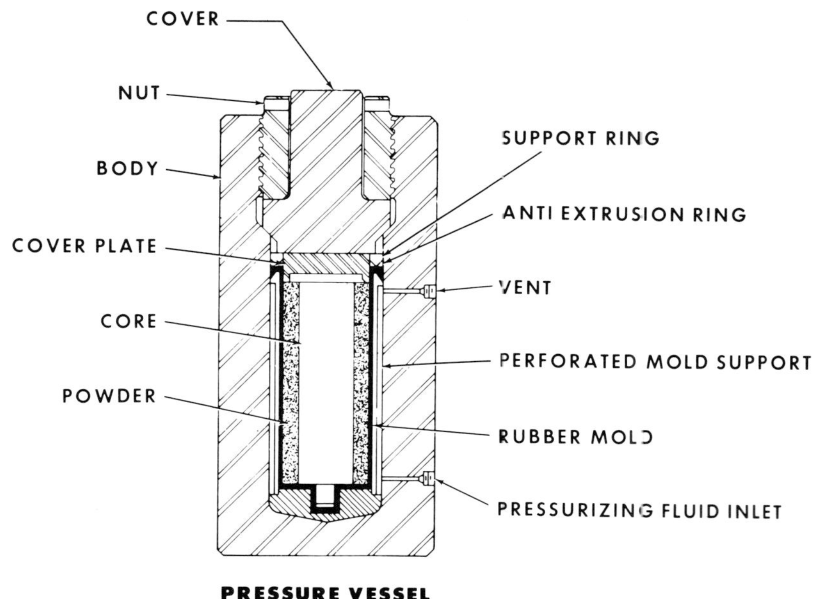

FIGURE 9B.
Pressure vessel with dry bag tooling

Undercuts can also be added if the largest dimension of the compacted part will clear the smallest dimension of the mold when withdrawn. Such advances as multi-chamber Fixed Mold Presses and automated powder fill and parts ejection mechanisms have made this process feasible for many mass production applications.

Automated systems are available for both Free Mold and Fixed Mold Isostatic Pressing. The selection of one process over the other is dependent upon the configuration of the part, production rates required, and cost considerations.

Pressure/density Relationship: Compared with die-compaction, isostatic compaction provides more uniform pressure distribution within the compact. the major reason being the absence of die-wall friction. Another factor is the greater area over which the pressure is applied. Furthermore, distinct advantages may often be gained from the ease with which air can be evacuated from the loose powder before compaction if necessary and because of starting with a uniformly packed powder. Consequently, isostatic compaction provides increased and more uniform density at a given compaction pressure, and relative freedom from compact defects when applied to brittle or to fine powders. It is a result of the uniform compaction pressure/density distribution that the cross-section: height ratio is not a limiting feature as it is with die-compaction. Another characteristic of the process is the elmination of die lubricants. Their absence permits higher pressed densities and eliminates problems associated with lubricant removal prior to or during final sintering.

Some typical compaction data for various materials are shown in the following Tables and curves.

Table 9.1
Comparison of die and isostatic compaction of metal powders

Material	Process	K Value, psi x 10^{-5}
Aluminium	Isostatic	3.3
Beryllium	Isostatic	1.0
Copper:		
(200 mesh)	Isostatic	12
Chemically pptd.	Die	1.3 - 1.9
Spherical	Die	3.0 - 3.8
Electrolytic	Isostatic	3.4
	Die	2.8
Iron:		
(−200 mesh)	Isostatic	1.1
Electrolytic (78% 65/325 mesh)	Isostatic	1.6
Electrolytic (90% 10-44 micro m)	Isostatic	2.3
Electrolytic (40-250 mesh)	Die	1.8 - 2.0
Hydrogen-reduced (140-200 mesh)	Die	1.5
Sponge	Isostatic	1.9
	Die	1.4
Tungsten:		
Coarse (40/100 mesh)	Isostatic	0.4
Fine (10 micro m)	Isostatic	0.8
(15 micro m)	Die	0.8

Table 9.2
Isostatic Compaction Data
Comparing Densities*

Pressure	Isostatic-ϕ R	Die-PR	Vibration	Acceleration
Ni-Powder				
150 Ksi - 10T/cm^2	77%	73%		
Cu-Powder				
20 Ksi - 1400 atm	7.80	6.80		
25 Ksi - 1800 atm	8.60	7.35		
30 Ksi - 2200 atm	8.80	7.60		
Fe-Powder				
20 Ksi - 1400 atm	6.00	6.00		
25 Ksi - 1800 atm	6.40	6.30		
30 Ksi - 2200 atm	6.90	6.70		
Tungsten-Powder				
20 Ksi - 1400 atm	11.30	10.80		
25 Ksi - 1800 atm	12.00	10.90		
30 Ksi - 2200 atm	12.40	11.00		
Co-Powder (with binder)				
30 Ksi - 2200 atm	65%	59%		
Al_1O_3-*Powder (<40μ)*				
15 Ksi - 1000 atm	*65%*			64%
ZrO_2 *(50-70μ)*				
15 Ksi - 1000 atm	*57%*			55%
ZrO_2 + 40%Fe (Cermet)				
15 Ksi - 1000 atm	72%			72%
$MoSi_2$ *(1-40 without binder)*				
40 Ksi - 2900 atm	*74%*	70%	63%	
85 Ksi - 6000 atm	76%	72%		
125 Ksi - 9000 atm	79%	75%		
MgO-Powder (500μ)				
15 Ksi - 1000 atm	64%		63%	61%

*Test samples were small (ϕ ¾ in. x length 1 in.). For larger samples, results would be more favorable to isostatic pressing.

The most important differences between the mechanisms of isostatics and die-pressing are:

I. Compaction forces and movements which act simultaneously from all sides result in less linear displacement of an important portion of the powder mass for obtaining comparable density levels. This results in somewhat less frictional work for each of the powder particles.

II. Existing pressures in and outside the powder environment are equal. Thus any trapped air within the powder mass cannot escape; so that it would be compacted between the powder particles, and would influence the pressure-volume relation of the powder mass.

Summarizing, the factors which define powder compaction are:

1. The particles can slide over each other without deformation to new equilibrium positions.
2. The particles can deform elastically, but only at and around the contact surfaces.
3. The particles can deform plastically.
4. The particles can deform by fracture deformation.
5. The mass of each particle may be compacted elastically due to applied pressure.

Among factors, the one which will predominate in the compaction mechanism will depend on the material itself, on the particle sizes and on the particle size distribution. It is evident that in all cases a combination of these factors will occur simultaneously or in a certain order of progression depending on the pressure level.

The pronounced change in slope of the pressure vs. density relationship, clearly shown in Figure 9C.1 for die-compacted powder, appears to occur also for isostatic compacting. Results for aluminum, beryllium, copper, and iron powders show such a change in gradient at high applied pressures which suggests that this is related to strain-hardening.

The relative densification obtained by die-and isostatic compaction is shown in Figure 9C.2 for several powders.

Generally, there is a distinct advantage in using isostatic conditions except for aluminum and iron compacted to high densities. At high densities both die and isostatic compaction produce similar green densities with iron and aluminum powders. For materials such as aluminum that have constant shear stress, the radial pressure becomes approximately equal to the axial pressure, i.e. approaches an isostatic pressure distribution. However, for materials such as copper where yield stress is a function of the normal stress on the shear plane, the radial pressure remains less than the axial pressure. Although the pressure distribution within a die-pressed compact may become isostatic, presumably the pressure vs. density relationship should be identical with that of isostatic compacting only if the density distribution is equally uniform.

Because with isostatic compaction the effects of die-wall friction are absent, much more uniform densities are obtained. This is in contrast to die-compaction where it is well established that die-wall friction exerts a major influence on density distribution in the absence of radial powder flow.

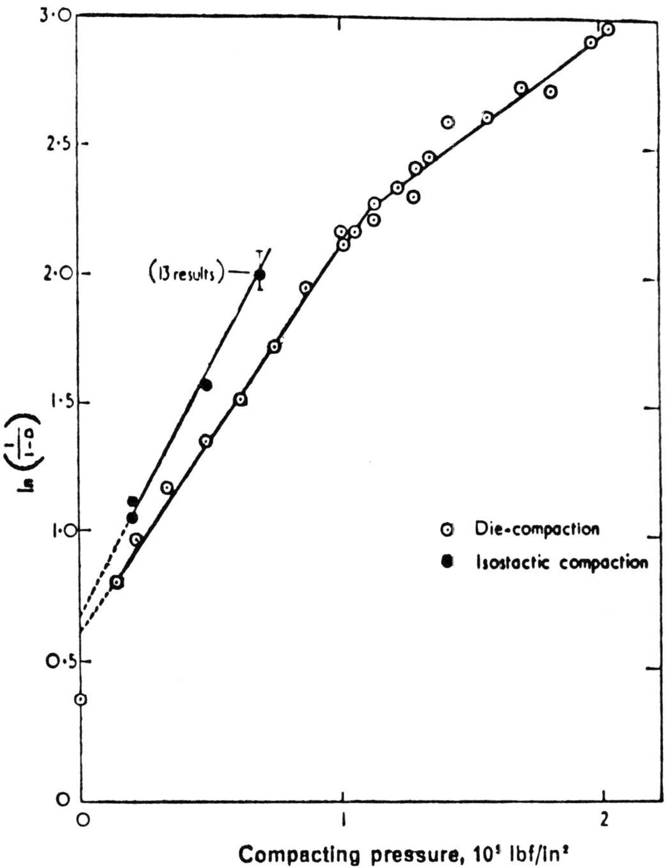

FIGURE 9C.1

A comparsion of the green density (D) of sponge-iron powder after die-or isostatic compaction

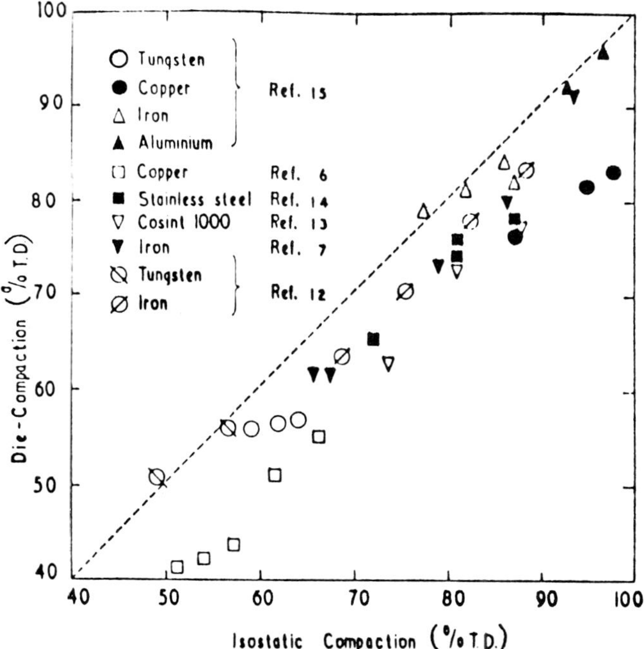

FIGURE 9C.2

Green densities achieved with various metal powders by die-or isostatic compaction at the same applied pressure (T.D.-theoretical density)

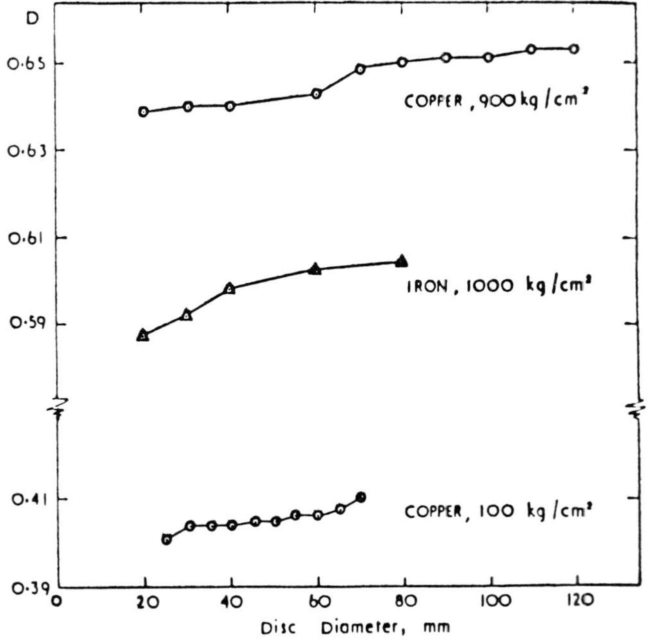

FIGURE 9C.3

The effect of compact size upon the green density (D) of isostatically compacted iron and copper powders (Borok)

9.2 TOOLING FOR ISOSTATIC PRESSING:

Isostatic pressing employs flexible tooling of neoprene rubber, urethane, polyvinyl chloride, or other elastomers as compared with tool steel and carbide dies of other P/M processes. Rubber is widely used, particularly when thin wall tooling is required. It is used also for pressing more complex shapes when more rigid tooling would present problems in extracting the pressed part from the mold, and in applications calling for disposable tooling. Where more rigid tooling is required, polyvinyl chloride and urethane should be specified.

Mold Parameters: Depending on shape size, dimensional requirements, quantity and the material itself, the molds for normal isostatic pressing, processing can be made by one of the following manufacturing techniques:

— Dipping
— Pressing (mechanical)
— Form casting
— Injection molding
— Autoclave curing

Commonly used materials are: Natural rubber, Neoprene, Urethane, certain plastics and reversible gel materials. The selection of a mold material depends on:

— compatibility with the powder being compacted
— compactibility with the pressure medium
— required compressibility depending on the compaction ratio of the powder
— required hardness necessary to resist wear by the powder material. This depends on the powder characteristics (rough and hard powder particles will wear out the mold material sooner) and the movements made during compaction depending on the compaction ratio and mold form

Hardness of the mold material can influence compressibility (elasticity) adversely and therefore is directly restricted to certain maximum values by the compaction ratio of the material to be compacted.

The mold design is influenced by factors such as:

a. The compaction ratio of the powder under isostatic pressure, its multidirectional influence and its proportion to form, dimensional proportions, and design details.
b. Shrinkage during sintering for parts obtained by isostatic pressing. This differs from the shrinkage obtained with parts densified by conventional methods.
c. Design factors necessary for a good fitting, sealing and eventual location of the mold or combined mold.
d. Dimension and form changes which happen in the curing process during manufacturing of the mold itself.

Consideration of all of these factors is necessary for the design of a proper mold in both form and dimensions. This can best be controlled by the press manufacturer. Therefore the fill form and fill dimensions should be clearly understood. The press manufacturer should also consider these factors when designing a suitable mold for high pressure work.

In addition to the elastometric mold itself, it may also be necessary to include auxiliary tooling elements such as:

— A mold support form to back-up the elastic mold with a solid support to help obtain exact form and dimensional control during filling.
— A vacuum form behind the support to pull the elastic mold exactly in place during each filling. This method is used mostly for the filling of free molds, if the parts are compacted with "green stage" tolerance requirements. For fixed mold processing the same action can be carried out by a reversed pumping action on the pressure medium.
— This vacuum form, or with fixed mold pressing, the reversed pumping action, can also open up the mold for stripping compacted parts that would normally not clear from the mold without widening.
— A vacuum connection to the fill area to de-air the filled mold before the compaction operation.

— A sealing support to back-up the sealing off of the mold cover or the mold before compaction.
— Clamping elements to carry out the sealing action.
— Mandrel or inserts for counter pressure.

Equipment Considerations: An important consideration in any Isostatic Pressing system is the design of the pressing equipment, and particularly the design of the pressure chamber. The pressure chamber must be designed to withstand the severe cyclic loading imposed by rapid production rates and must take into account fatigue as a significant mode of failure.

Equally important is the design of the cover or closure. Not only must it reliably withstand high pressures and cycling conditions, but it must also be designed for quick, easy opening and closing. Several types of closure designs are currently being used to meet these requirements, including threaded, pin and hydraulic clamp types. These are available both in manual and automatic systems.

Production Speeds: Production speeds are largely dependent on the size of the desired compact. The closure, the degree of automation in powder filling, pressurization and parts ejection operations, and the speed and capacity of the pressurization system all affect the overall speed. Standard automatic Isostatic Presses are currently being cycled in ten seconds or less for small parts using single or mulitple cavity tooling and 120 seconds or less for parts weighing over 1,000 pounds. Current presses offer continuous automatic production from powder fill to parts ejection.

Pressures being used now range to 150,000 psi, although this extreme is limited to research or limited production applications. Units presently are being designed with inside diameters to 60 inches at lower pressures (10 ksi to 30 ksi), and diameters to 24 inches at higher pressures (80 ksi to 100 ksi).

Through the use of automated equipment and careful processing including particle size control, avoiding contamination of the powders, maintaining uniform temperature and humidity as well as pressure and time, excellent dimensional control can be achieved.

Table 9.3 Typical pressures required for isostatic pressing of powders

Material	Pressure
Teflon:	approximately 2,000 psi to 10,000 psi
Plastic:	approximately 4,000 psi to 10,000 psi
Ceramic:	approximately 8,000 psi to 20,000 psi
Aluminum:	approximately 8,000 psi to 20,000 psi
Iron:	approximately 45,000 psi to 60,000 psi
Stainless steel:	approximately 45,000 psi to 60,000 psi
Copper:	approximately 20,000 psi to 40,000 psi
Lead:	approximately 20,000 psi to 30,000 psi
Tungsten Carbide:	approximately 20,000 psi to 30,000 psi

These pressures are approximate since every application presents its own unique requirements of density, configuration and size.

9.3 HOT ISOSTATIC PRESSING:

Hot isostatic pressing (HIP) is a process which only superficially resembles cold isostatic compaction. Whereas in cold isostatic systems, the compaction phenomenon is basically mechanical, with HIP there are significant metallurgical effects taking place such as solid state diffusion. In principle, the powder is placed inside a sealed mold and then is subjected to isostatic pressure at any elevated temperature, in excess of 538°C (1000°F) and usually at 982°C (1800°F), 1260°C (2300°F), 1427°C (2600°F) or 1649°C (3000°F) depending upon the product being processed. Temperatures are usually selected to avoid any liquid phase although in some complex alloys or mixtures a partial liquid phase in unavoidable, particularly in binder materials. It is because of the elevated temperatures that procedures and techniques vary from that of cold processes.

Because rubber or other conventional elastomers cannot withstand the imposed heat, metal or glass are used as tooling. This automatically restricts shape and configuration to relatively simple forms. Thus the process is normally carried out in the free mold manner. The reasons for this will become evident when the typical cycle is described.

Tooling is of the isostatic type, although some experimental work has been done to produce complex configurations using a quasi-isostatic tooling system wherein a high temperature ceramic such as alumina is used as the rigid member.

Typical Cycle:

1. The powder to be compacted is placed inside a container or mold, usually made from mild steel. The filled can is then closed by welding a top to it fitted with a suitable evacuation connection.

2. The evacuation and outgassing of the powder is usually carried out at a moderately elevated temperature. A suitable level is 427-649°C (800-1200°F) for alloy steels and super alloys. This is accomplished by connecting a vacuum pump to the evacuation connection which is extended to reach outside the preheating furnace. The container is pumped down until a level of at least 1,000 microns is reached and the work is then placed inside a preheat furnace and brought up to temperature. Once temperature is reached, vacuum pumping is continued until the 1,000 micron level is reached inside the container. The work is then removed from the furnace, still under vacuum and the connection tube is crimped and welded shut close to the can and the excess length trimmed off to permit efficient loading into the hot isostatic press. These two steps can be done independently of the HIP unit so suitable loads can be prepared in advance and stockpiled to keep the HIP unit efficiently utilized.

3. One or more prepared cans of material are loaded into the work space of the HIP unit. The number and arrangement are a function of the geometry involved, modified only by the necessity to insure that the desired temperature uniformly be maintained and that adequate mechanical support is provided. Conventional alumina kiln furniture is used for some parts. In other cases crucibles and/or boats can handle small parts. Various grades of alumina powders and balls can also provide the required mechanical support. It is common to bench pack a large mild steel cylinder, which is sized to fill the work space of the system with as many parts as feasible so that only a single element need by actually inserted and removed from the press. In small laboratory size equipment, loading is done directly into the HIP unit, but the method described previously is used on all larger systems.

4. The HIP unit is closed and locked, evacuated to minimize contamination of the pressurizing gas, and then the pressure and temperature are usually raised simultaneously. In some cases; e.g. use of glass as a canning material, minimum temperature levels are established first, followed by pressurization to insure that the canning material used is adequately ductile. In other cases, pressurization to an intermediate level is done cold and then the pressure is raised to the designed working level by increasing the temperature.

5. After reaching the desired pressure and temperature levels, the system is kept at a "hold" or soak condition to insure that the center of the load has reached the set conditions. No generalizations can be made covering the time required since it is totally a function of the load characteristics, shape, mass, composition, configuration of loading, etc. Thermocouples placed in the load indicate completion of the minimum soak time. Additional time might be specified to achieve particular metallurgical properties. These are separately

evaluated on the basis of diffusion rates and similar factors in a given application.

6. Finally, pressure and temperature are allowed to drop. Normally the furnace is merely shut off completely and the temperature is allowed to decrease at its maximum rate. Pressure is usually allowed to drop in thermodynamic reaction to the temperature drop since bleeding off gas to reduce the pressure will slow down the cooling rate. Some materials require a slower, controlled cooling which can, of course, be achieved in most HIP systems. Cycles may be timed so that the cool down phase can take place at night. In the morning the system is ready for venting to zero pressure and reloading. This is a normal production cycle condition. There are some operations which can achieve two cycles per day, while others, by the nature of the cycle required by the product, might run up to two or three days.

7. The vessel is then opened, reloaded and the above steps repeated.

Reference

Figures 9A, 9C.1, 9C.2, 9C.3 and Table 9.2 are used with the permission of Interceram Magazine, published by Verlag Schmid GmbH, Freiburg, Germany. Some additional data are used with the permission of the Metals Society, London, England.

PART
FOUR

SINTERING

10.0 FURNACES FOR ANNEALING AND REDUCING METAL POWDERS

Large continuous conveyor type furnaces are used to anneal most of molding grade metal powders manufactured today. These furnaces also reduce the oxygen to an acceptable commercial level. Iron and copper are the two major powders processed.

A continuous stainless steel strip conveyor belt carries powder through the furnace. A protective atmosphere, normally dissociated ammonia or hydrogen, is used.

FIGURE 10A.
Gas-fired radiant-tube heated iron powder reduction and annealing furnace

The smaller furnaces for lower tonnage production can be muffle-type furnaces. Larger tonnages use roller-hearth conveyor furnaces, in which the continuous conveyor is carried on a roller hearth (fig. 10A).

These furnaces may be heated either by gas or conventional electrical means, although the majority of equipment in operation today is gas fired.

11.0 FURNACES FOR SINTERING-PRINCIPLES OF CONSTRUCTION

Sintering furnaces used for production of powder metallurgy products are somewhat similar to furnaces used for copper brazing of steel. However, there are certain fundamental principles of design that must be incorporated into a furnace for sintering that may not be required for brazing.

Generally speaking, in brazing, parts can be brought up to temperature rapidly, and need to be held only long enough for the copper or other brazing material to melt and flow. Parts to be sintered must be brought up to temperature at a slower and controlled rate of heating, and are then held for a soak period at the sintering temperature to obtain the desired density. Furthermore, a "burn-off" or heated purge chamber is essential to expel the air and the lubricant entrapped in the voids of the green compact before it entèrs the sintering chamber. This is necessary to reduce contamination of the furnace atmosphere. If the volatilized hydrocarbons and zinc from the lubricants are not expelled before the compacts reach the high temperature portion of the furnace, carbon and zinc oxide will deposit on the heating elements and

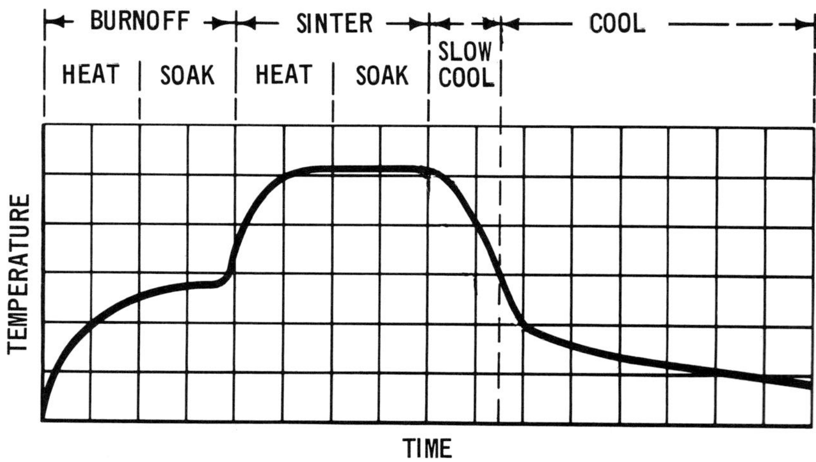

FIGURE 11A.
Typical temperature profile in a continuous sintering furnace

furnace refractories, causing high furnace maintenance. The carbon will also cause discoloration of the compacts. Furthermore, if the compact is heated suddenly, the escaping lubricant may rupture the compact or affect its dimensions and make it difficult to control the tolerances of the part.

Construction of a sintering furnace requires (a) a gas-tight furnace shell or a furnace containing a gas-tight muffle to maintain a reducing furnace atmosphere, (b) a heated burn-off or purge chamber to expel the air and lubricant vapors, (c) a controlled rate of pre-heat, (d) a controlled high-heat sintering chamber, and (e) a water-jacketed cooling chamber, with a protective atmosphere to prevent oxidation while the compacts cool. Figure 11A shows a typical temperature profile in a continuous sintering furnace.

11.1 BURN-OFF AND ENTRANCE ZONE:

The burn-off chamber, which is sometimes referred to as the heated purge chamber or entrance zone, is a very essential part of the sintering furnace. The common lubricants used in compacts are stearic acid, or various metallic stearates. These compounds are usually added to the powder to provide die lubrication and to allow the powders to flow more easily under pressure. The lubricants are of no value in obtaining good bonding and sintering. In fact, the lubricants will interfere with good sintering if not expelled properly before the compacts reach the high temperature portion of the furnace. The compacts should be heated slowly at this stage so that the entrapped air and lubricants will not expand too rapidly and push the metal particles apart.

It is very important that all of the lubricants be volatilized and expelled by the outflowing furnace atmosphere before the compacts enter the high temperature zone of the furnace. The burn-off chambers are usually controlled to heat the compacts to approximately 800° F to 1500° F (430° C - 816° C). Sufficient input should be provided to heat to at least 1600° F (870° C), if desired. Different types of lubricants volatilize at different temperatures. The burn-off chamber must also be of sufficient length to allow the compacts to reach the temperature needed to completely volatilize the lubricant. A common error, where varying sintering cycles are used, is to have the burn-off chamber designed for the longest sintering cycle to conserve floor space and cost, instead of designing for the shortest cycle, so that ample time is allowed in the burn-off chamber for all cycles. The heating rate is the slowest in the low temperature range, and this factor is often overlooked.

If the lubricants are not expelled completely before the compacts enter the high temperature sintering zone, considerable difficulty will be encountered. Since the furnace protective atmosphere is reducing, there is no air in the furnace to burn up the hydrocarbon vapors. The hydrocarbons break down into free carbon which deposits on the refractories and on the electrical heating elements. If the furnace is not shut down frequently to burn out the carbon by means of introducing air into the furnace, the heating elements will short out and fail, and the refractory will fail due to impregnation with carbon. Even if the furnace is of the full muffle type, the carbon will deposit on the muffle and cause poor heat transfer, which will lead to premature failure.

When zinc stearate is used as a lubricant, the zinc as well as the stearate hydrocarbon will add to the problem of operating the furnace, if not expelled completely in the burn-off chamber. In this case, the zinc will volatilize in the high temperature zone and will contaminate the heating elements and the furnace alloy components in an atmosphere-tight furnace. Eventually it will work its way to the cooling chamber and condense on the sides. In the cooling chamber this will reduce the heat transfer and the compacts will not cool sufficiently. If this condition exists, the furnace must be shut down, and the cooling chamber scraped out.

Not only will the lubricants cause excessive furnace maintenance if not completely expelled before the compact enters the high temperature zone, but they can affect the quality of the product. The dimensions of the parts may vary because the lubricants are being expelled too rapidly when the parts reach the heat zone. Carbon from the lubricants may discolor the part and make it a reject. The discoloration can vary from part to part, because the lubricants may have been completely expelled from some and not from others.

The flow of the atmosphere also plays an important part in expelling the lubricant vapors when the burn-off chamber is attached to the high heat sintering furnace. Sufficient atmosphere gas must be provided, and the flow directed so the vapors are discharged and do not flow into the high heat zone. This is not difficult to do in properly designed sintering furnaces.

Another way is to separate the burn-off chamber from the high heat sintering chamber. The two chambers are separated with an air gap, utilizing a flame curtain before the high heat chamber. Thus, any volatiles escaping from the burn-off chamber cannot enter the high heat chamber when a load of compacts is charged, provided the compacts have been heated sufficiently to drive off all the lubricant. The separate burn-off chambers are usually semi-muffle gas-fired furnaces. They will burn up the volatile lubricants and discharge them with the flue products.

The separate burn-off chambers are usually operated so the compacts reach a maximum temperature of 800° F (430° C). The products of combustion in general do not excessively oxidize the green compacts up to this temperature. Combustion can be adjusted toward the reducing side to keep oxidation to a minimum. However, when processing compacts with graphite additions, this type of burn-off chamber can lead to partial or total loss of graphite due to oxidation.

A considerable number of gas-fired, semi-muffle separate burn-off chambers have been installed on continuous furnace installations, and excellent sintering has been obtained with the complete absence of high furnace maintenance associated with these furnaces when using attached burn-off chambers.

If the burn-off chamber is attached to the sintering furnace, the design of the burn-off chamber's electric heating elements must be such that the hydrocarbon vapors, zinc and carbon do not affect them. This is done by either using a muffle at this point, or by using sheathed elements or by using a low voltage element. As a substitute for electric elements, gas fired radiant tube heating units can be used.

Considerable attention should be given to the design of the burn-off chamber because of its importance to good sintering and low furnace maintenance, and the fact that it also serves as a first stage preheat.

Burn-off of the lubricants for vacuum sintering must be given special attention. In the first place, the metallic stearates must not be used. Instead, stearic acid or a wax compound should be used. Zinc or lithium residue left behind after the burn-off of a metallic stearate will volatilize at the high sintering temperature and contaminate the furnace and vacuum pumps. For batch-type vacuum furnaces, the burn-off is best accomplished in a separate protective-atmosphere oven at about 800° F (430° C). An air atmosphere in an electric oven or the products of combustion in a gas-fired oven will degraphitize iron-graphite P/M parts at temperatures as low as 700° F (370° C). This will result in a decarburized layer after sintering in a vacuum furnace. Air-atmosphere electric ovens are quite suitable for dewaxing the austenitic stainless steel P/M parts at about 800° F (370° C). The carbon-bearing martensitic stainless steels should be dewaxed in a protective atmosphere to prevent decarburization before vacuum sintering.

Special semi-continuous furnaces have been built for the sintering of tungsten carbide P/M compacts, in which the carbide parts are dewaxed under vacuum. A cold trap is used to condense and separate the volatilized wax before reaching the vacuum pumps. The vacuum dewaxing furnace is designed so that the walls of the furnace are above the melting point of the wax to prevent condensation of the wax on the furnace walls or heat shields. Vacuum dewaxing has several advantages over protective-atmosphere dewaxing: first, the dewaxing takes place at a much faster rate, and second, vacuum is less costly than protective atmosphere and is more thorough in the removal of the wax at the low temperature.

11.2 HIGH-TEMPERATURE ZONE:

The high-heat chamber must be of the proper length in relation to the burn-off and/or preheat zones to allow sufficient time at temperature to obtain the desired density and strength. The chief causes of poor strength or density in a powder metallurgy product is that either a sufficient temperature was not reached, or that the part was not held long enough at the proper temperature to obtain good bonding of the particles.

At the outset the statement was made that a furnace designed for brazing may not make a good sintering furnace. To illustrate: a furnace designed to copper braze or heat 500 pounds of steel parts per hour to 2050° F (1120° C), might sinter only 250 pounds or less of iron parts per hour, using the same loading density. This is because wrought steel to be brazed need be heated to only 2050° F (1120° C) to melt the copper, but the iron parts to be sintered also must be soaked at 2050° F (1120° C), and thus the production rate is decreased by 50% or more. Therefore, in sintering furnaces, the high-heat chambers are made longer, and the heat input is concentrated at the entrance end and reduced in the soaking zone.

11.3 COOLING ZONE:

The cooling zone of a sintering furnace often consists of a short insulated cooling zone, followed by a long water-jacketed cooling zone to cool the parts and prevent oxidation upon discharge into air. The short, insulated cooling zone permits the parts to cool from the high sintering temperature to a lower temperature at a slower rate so as to prevent thermal shock. The insulated cooling zone also cuts down on maintenance of the water-jacketed cooling zones, belts, trays, and fixtures by preventing the high stresses from thermal shock. The temperature to which the parts must be cooled before discharge depends on the material being sintered.

Furnaces become quite long if cooling is to be provided below 300° F (150° C). The cooling rate is exceedingly slow at the low temperature range. Fans sometimes are employed in the cooling chambers to help cool by circulating the atmosphere over the work. Extreme care must be taken in the design when employing fans, so that air is not sucked into the cooling chamber when the doors are opened.

Automatic water temperature control is essential on cooling chambers for fool-proof operation. If a cooling chamber is operated too cool, below the dew point of the furnace atmosphere, condensation will occur on the walls of the chamber, and the iron parts will become blue or oxidized. If the chamber is operated too hot, the parts will not cool sufficiently, and will come out of the cooling chamber hot and oxidize in the air. To compensate automatically for varying loads, automatic flow control of the water by means of a thermostat and throttling control valve is the only fool-proof method of operation.

11.4 MUFFLE vs. GAS-TIGHT SHELL CONSTRUCTION:

In view of the fact that much of the pilot plant work and laboratory work in sintering is carried out in small muffle furnaces, many question the application of gas-tight shell construction, non-muffle electric or radiant tube gas-fired furnaces when going to large production operation. Generally, muffle construction should be eliminated from large production sintering furnaces, unless there is an absolute necessity for very close control of the furnace atmosphere purity. Large muffles are not only expensive initially, but are also a high maintenance item. Heating through a muffle is less efficient than having the electric heating elements exposed to the furnace atmosphere and furnace chamber. The general rule for large production furnaces is to omit the muffle when using exothermic or endothermic atmospheres. If the furnace is to be direct gas fired, then the use of a full muffle is a must to prevent oxidization of the compacts.

Full-muffle construction or a high purity refractory lining is necessary when employing hydrogen or dissociated ammonia atmospheres and when dew points of less than -20° F (-29° C) are required. The successful sintering of stainless steel requires the dew point of the hydrogen to be less than -20° F (-29° C) at 2100° to 2200° F (1150° - 1204° C). A full-muffle furnace must be used for this application because the hydrogen will react with the silica in the refractories at these temperatures to produce water vapor and raise the dew point. An exception to this is the application of very high purity aluminum-oxide refractories used in molybdenum element furnaces, to be described later in more detail. These special refractories are extremely expensive, and the furnace must be operated continuously to obtain and maintain a low dew point atmosphere.

Another advantage of a full-muffle furnace when using hydrogen or dissociated ammonia atmosphere, is that the flow of atmosphere purging gas can be reduced considerably over the larger non-muffle chamber. Since hydrogen and dissociated ammonia atmospheres are relatively expensive, this is an important consideration. Full-muffle furnaces purge down fast because there is no entrapped air to be removed from the brickwork. One method of purging full-muffle furnaces is to purge with dry nitrogen or some other inert gas, so that the hydrogen will not burn to form water vapor when introduced into the muffle. The nitrogen also prevents the muffle from oxidizing and then the oxides being converted to water vapor to contaminate the hydrogen atmosphere when it is introduced into the hot furnace muffle.

Another method of purging full-muffle furnaces is to bring the furnace up to a temperature of 1400° F (760° C) before hydrogen or dissociated ammonia are admitted. The oxides formed in the muffle due to lack of atmosphere gases up to 1400° F (760° C) and the moisture content first developed by the atmosphere gases, can be eliminated during the time that the furnace is being heated up from 1400° F (760° C) to the normal operating temperature.

On small full-muffle furnaces, the maintenance is not too great even at temperatures as high as 2100° F (1150° C) and, therefore, small furnaces are made full-muffle to take advantage of the fast purge down for intermittent use, and also to be universally adapted for use with hydrogen, exothermic, or endothermic atmosphere.

11.5 TYPICAL PROTECTIVE ATMOSPHERES USED FOR SINTERING:

The most widely used protective atmospheres for sintering are: hydrogen, dissociated ammonia, rich exothermic gas, purified rich exothermic gas, dry or wet endothermic gas, and vacuum. The factors involved in the selection of the proper atmosphere gas tie in with the appearance, properties, and cost of the sintered compacts. Atmosphere gases are discussed in considerable detail under "Furnace Atmospheres and Atmosphere Gas Producers" in section 19.0.

All five of the above-mentioned gases and vacuum have properties which help to reduce oxides in powder particles and promote maximum sintering. They also prevent oxidation and thereby reduce interparticle friction while assuring good machinability and good surface appearance.

Atmospheres, such as purified, rich exothermic gas, dry endothermic gas and vacuum are desirable for carbon-steel compacts. They tend to minimize decarburization in the surfaces or bodies of the parts which would result in soft wearing surfaces and low physical properties after heat treatment, and render the parts useless.

Once the most suitable protective atmosphere has been selected, the quality of the protective atmosphere must be maintained for optimum mechanical and physical properties, and minimum discoloration and decarburization of the work pieces.

Nonferrous compacts, such as copper and bronze, may be oxidized throughout, and scaled or discolored by oxygen. They are not adversely affected by hydrogen and carbon monoxide.

Brass compacts are adversely affected by carbon dioxide as well as oxygen, sulphur and water vapor, due to selective attack on zinc. This impairs mechanical and physical properities, as well as appearance. When sintered in an open furnace, brass compacts generally are treated only in pure, dry hydrogen or dissociated ammonia. Vacuum is not suitable for sintering brass because the zinc will volatilize.

Oxidation of compacts made of iron or iron-graphite mixes is caused by oxygen, water vapor, and carbon dioxide. If present in unsatisfactory proportions with respect to hydrogen and carbon monoxide, they can cause discoloration or scaling. Iron oxides are reduced by hydrogen, carbon monoxide and carbon. Decarburization is caused by oxygen, water vapor, and carbon dioxide. Carburization, on the other hand, is caused by carbon monoxide and hydrocarbons such as methane. Vacuum sintering will neither carburize nor decarburize iron-graphite mixes.

Prerequisites of a sintering furnace: The fundamental principles of design of a sintering furnace have been discussed, but the final choice of the type of furnace depends upon many factors, such as the sintering temperature, the material to be sintered, the sintering atmosphere, and the production, operating and maintenance costs for a given production. In order to make an intelligent choice of a sintering furnace for a specific application, some other factors must be considered. These include: the principles of fuel-fired furnaces and their economies; the principles of electric furnaces; the effects of furnace atmosphere on refractories; methods of conveying compacts through furnaces; furnaces for heat treatments other than sintering; and finally, temperature control and pyrometry in sintering furnaces.

12.0 FUEL-FIRED FURNACES

Fuel-fired furnaces are either direct-fired muffle, or radiant-tube-fired. The reason for this, of course, is the necessity for isolating the work and its protective atmosphere from the products of combustion. Limited use is made of semi-muffle furnaces and at pre-sintering temperatures only where it is desired to remove lubricants from the compacts. Fuel-fired furnaces are commonly applied to the sintering process where fuel economy and low-range temperatures are dominant factors.

Uniformity of heating and close control of atmosphere conditions are critical factors in the sintering of metal powders. Both the direct-fired muffle and radiant-tube-fired furnaces meet these requirements.

12.1 MUFFLE-TYPE:

One of the prime considerations in the use of direct-fired muffle furnaces is the proper type of furnace muffle to be used. High temperature operation requires a material to withstand these temperatures, and in addition be capable of withstanding the thermal shock which results from pushing the work through the furnace. Refractories fulfill the temperature requirements, but the thermal conductivity and thermal shock properties of refractories are frequently not adequate. Also, cracks may develop rather easily, joints may not be tight, and the refractory may be porous. These factors can allow products of combustion and/or air to leak in, but cannot be tolerated since they will contaminate the atmosphere. Therefore the use of metallic alloy muffles normally is a definite requirement.

Since very dry high-purity atmosphere is a requirement for sintering stainless-steel parts, metallic muffles are almost universally used in this application. The most popular shapes of muffles have "D" or circular sections, often supported fully in their lengths by refractory hearth plates. A furnace with cylindrical muffle on refractory piers is shown in figure 12A.

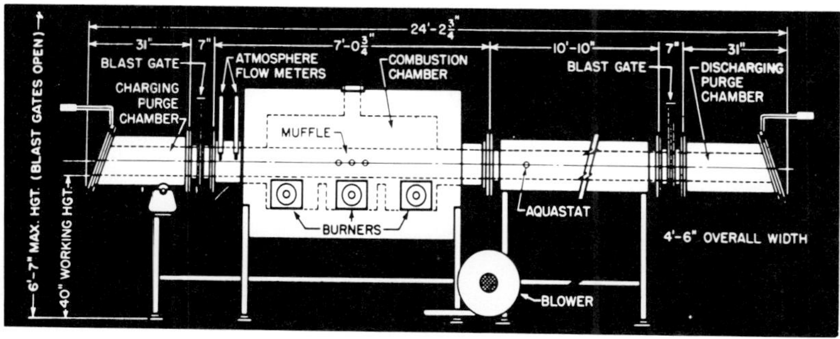

FIGURE 12A.

Principles of construction of gas-fired, full muffle furnace for maintaining low dew point control of furnace atmosphere

Muffles require the proper constructional strength to minimize sagging, bulging, warping, and cracking at high temperatures.

In order to insure good atmosphere conditions during normal operation and oxide-free surface on the inside of the muffle, a reducing atmosphere must be maintained when the muffle reaches the red heat range. On shut downs, the atmosphere must be maintained until the black heat range is reached.

Muffle furnaces are most satisfactory for low production service, and where ultra-high purity and conservation of atmosphere is an important factor.

12.2 RADIANT-TUBE TYPE:

Radiant tubes are used to heat high-production sintering furnaces. These furnaces do not require an alloy muffle, but have an atmosphere-tight refractory-lined casing. They have the advantage of less maintenance and conservation of alloy. A typical cross-section of a vertically-fired radiant-tube furnace is shown in figure 12B. Much sintering work can be done in atmospheres that are not of the highest purity type, and the atmosphere-tight

FIGURE 12B.

Cross-section of radiant tube furnace showing how vertical tubes can be utilized for single row or multiple row furnaces

radiant-tube-fired furnaces satisfactorily meet these requirements. Although better purging and less atmosphere are advantages of the muffle furnace, furnace size and type limitations restrict muffle usage. Radiant-tube-fired furnaces permit low maintenance operation up to 1850° F (1010° C) when using nickel-chromium alloy tubes, and up to 2000° F (1093° C) when using super alloys or ceramic-coated tubes. Because of the temperature limitation, the most common applications of radiant tubes in sintering furnaces are in burn-off chambers.

12.3 ECONOMY OF FUEL-FIRED FURNACES:

Economical operation of the horizontally- and vertically-fired radiant-tube-furnaces is limited to operations below 1850° F (1010° C). Above this temperature, stack losses are excessive. See Table 12.1.

TABLE 12.1

Exhaust Gas Temp. °F	°C	Stack Loss %	Available Heat % Gross Input
400	(204)	16	84
600	(316)	21	79
800	(427)	26	74
1000	(538)	30	70
1200	(649)	35	65
1400	(760)	40	60
1600	(871)	45	55
1800	(982)	50	50
2000	(1093)	56	44
2400	(1316)	67	33
2800	(1538)	80	20

Also, operation above 1850° F (1010° C) leads to high maintenance costs of the alloy radiant tubes. Because of this, electric furnaces generally are used for high-temperature sintering. An exception is noted in the case of the small diameter alloy muffle furnaces which can be operated successfully up to 2100° F (1149° C), but this should not be expected on the larger furnaces where fuel economy is more significant.

12.4 SEMI-MUFFLE TYPE FURNACES:

The work chamber of fuel-fired semi-muffle furnaces is not isolated from the products of combustion, and no attempt is made to use atmosphere in these furnaces. The muffle protects the furnace load from direct flame impingement; however, the products of combustion can and do enter the muffle.

This type of furnace is used to burn off the volatile matter in the compacts prior to sintering. To avoid or minimize oxidation, the furnace is used to heat the work to 800° F (427° C) maximum. The products of combustion and the volatiles are discharged together through the furnace vent stack.

Summary: When determining whether to use fuel-fired furnaces or electric furnaces, fuel costs are not the only factor to consider. If a general statement could be made regarding a choice, it might be this: the choice of a fuel fired furnace or an electrically heated furnace should be predicated on the cost and availability of alternate heating sources. However, burn-off chambers can be either gas-fired or electrically heated. To select a gas-fired furnace merely because gas happens to be cheap in the particular area of the furnace operation might be unwise, unless engineering calculations have been made to determine the efficiency of heat release through the muffle to the work. As noted previously and shown by the heat-loss table, the stack losses mount rapidly with increasing temperature of operation. The restriction of heat release through the muffle or radiant tube to the work is another big factor.

When using high-purity hydrogen and dissociated ammonia in small gas-fired furnaces, alloy-muffle types are most successful in maintaining the high-purity atmosphere free from contamination. Muffles sometimes are employed in electric furnaces where extremely dry hydrogen atmospheres are required.

13.0 ELECTRICALLY-HEATED FURNACES

There are three basic kinds of electric heating elements used in sintering furnaces: (a) base-metal nickel-chromium alloys, (b) non-metallic heating elements, silicon-carbide or graphite, and (c) refractory-metal heating elements, molybdenum or tungsten. Graphite and refractory metal heating elements are used almost exclusively in vacuum furnaces.

All three types have definite and specific application to sintering. Like any process, there can be some overlapping in the application, but the wrong application, particularly with reference to furnace atmosphere, can cause very high furnace maintenance. Since the furnace atmosphere is so important to the life of the heating elements, this subject will be dealt with in more detail later.

13.1 NICKEL-CHROMIUM ALLOY HEATING ELEMENTS:

Nickel-chromium alloy heating elements are employed in electric furnaces for heavy duty continuous operation up to 2100° F (1149° C) in any type of furnace atmosphere except strongly carburizing or sooting atmospheres. The 80% nickel-20% chromium alloy generally is employed for the higher temperature ranges. The 35% nickel-20% chromium alloy is suitable up to 1800° F (982° C) operating temperature in protective atmosphere. There are some cases of contamination of furnace atmospheres, to be discussed later, where the 35% nickel-20% chromium alloy is more suitable than the 80% nickel-20% chromium alloy.

Heating elements are generally designed in solid rod overbend or flat ribbon overbend loops, and are supported on the refractory walls of the furnace by means of alloy pins, hanger bolts, or refractory supports as shown in figure 13A. Also, they can be mounted in the furnace hearth and supported from the furnace roof or arch. Elements can be used to distribute the heat evenly, or concentrated in areas where most needed. Some element mountings are built-in, while others are removable. The elements are strong and ductile when either cold or hot.

The watt density loading, i.e. watts per square inch of element surface, plays a very important part in the life of the alloy heating element. The higher the operating temperature, the lower should be the watt density loading. For example, if the heating element is to operate at temperatures in excess of 1750° F (954° C), the watt density should be in the order of 8 to 12 watts per square inch of surface area (1.24 watts/cm^2 to 1.86 watts/cm^2). For temperatures under 1750° F (954° C), the watt density may go as high as 17 watts per square inch (2.64 watts/cm^2), depending on the operating temperature and whether the furnace is to operate continuously with heavy production loads, or intermittently with light production loads.

FIGURE 13A.

Method of supporting overbend heating elements on a furnace wall

The cross-section of the alloy plays a very important part in the life of the heating element. Since most reducing atmospheres adversely affect the life of the element, the cross-section of the alloy should be as heavy as practical for a given design and input. The reason for this is that the attack by the atmosphere at the surface will cause a great change in the resistance of a thin cross-section, but will have only a small percentage of change on a heavy cross-section in the same period of time.

Heavy-cross-section heating elements not only cost more than the thin-section elements, but also may require transformers because the heating element voltage often is well below the normal 230 or 460 volt supply. This extra initial expense is a good investment, because the maintenance expense when using the heavier-cross-section elements will be reduced greatly.

Another good reason for operating heating elements of any type on low voltage is to minimize shorting and arcing out between the element bends and leads, due to carbon deposition from the atmosphere, or lubricants that may carry over into the high-temperature sintering chamber. Under favorable conditions of design, atmosphere, temperature and care, nickel-chromium elements have long life and low maintenance cost.

13.2 NON-METALLIC HEATING ELEMENTS:

(a) Silicon Carbide: Non-metallic silicon carbide heating elements are made in rod form. The ends of the rods extending through the insulation of the furnace have lower reistance, so that only the portion in the heating chamber gets hot. The rods usually run horizontally under and over the hearth as shown in figure 13B. However, it is also possible to run the rods vertically. The electrical connection is made by aluminum braid fastened to the rod by a

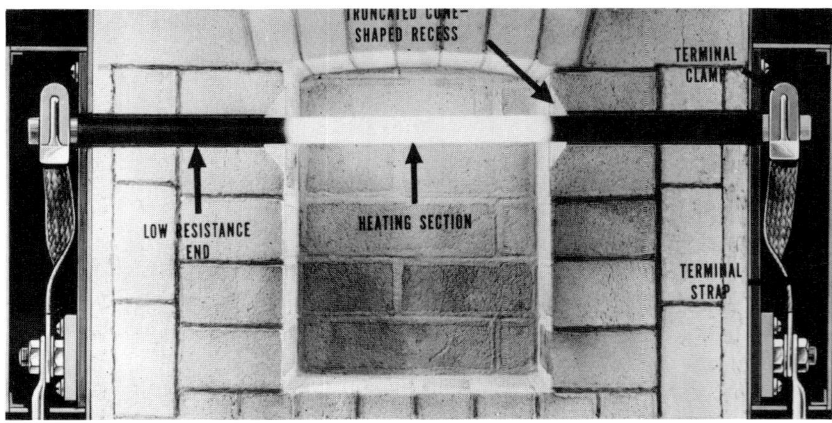

FIGURE 13B.

Method of installing non-metallic, silicon carbide heating elements in a furnace

chromium-steel spring clip, and the other end of the braid bolted to the bus bar, as shown. The terminal box on an atmosphere furnace is sealed with a cover plate and gasket. The silicon-carbide bars have low tensile strength when cold or hot and cannot withstand shock or high stress. They must be able to expand and contract freely due to temperature changes.

These elements can be replaced quickly and easily. In fact, the bars can be replaced while the furnace is hot. This is an important factor in large continuous furnaces which would require days to cool to get inside and replace built-in heating elements.

This type of element is particularly useful in sintering furnaces operating from 1850° F to 2100° F (1010° to 1150° C) in non-decarburizing atmospheres and in plants which require a furnace to be universally used with high or low carbon potential atmospheres on both ferrous and nonferrous parts. Reducing atmospheres adversely affect their life above, but not below, 2425° F (1330° C) element temperature, or about 2350° F (1290° C) furnace temperature.

A conservative watt density loading on silicon-carbide elements is 30 watts per square inch (4.65 watts/cm^2) in reducing atmospheres at 2100° F (1150° C). Higher loadings may be used when the elements operate in air. The high watt density loading of the silicon-carbide element is another advantage over the nickel-chromium type. More power input can be concentrated in the furnace and, therefore, reduce the wall area and overall size.

Silicon-carbide elements increase in resistance on aging, and variable-voltage transformers or saturable core reactors are normally required. This type of transformer is more costly than the constant-voltage type used with nickel-chromium elements. This is one disadvantage of silicon-carbide resistors and usually the initial cost of the overall equipment installation is greater because of it. Other disadvantages are that they do not all age at the same rate, particularly if connected in series-parallel circuits, necessitating matching resistance upon replacement. They are brittle and fragile at all temperatures, and show higher thermal losses due to the terminal holes in the walls of the furnace. However, the advantages more than outweigh the disadvantages, and silicon-carbide elements are widely used in sintering furnaces.

(b) Graphite: Graphite heating elements in the form of tubes, rods, plates, or cloth are used extensively in vacuum sintering furnaces. Since vacuum furnaces operate on the absence of air or oxidizing gases, graphite is a suitable heating element material up to 4000° F (2222° C). Graphite is relatively low in cost and is the only material that increases in strength as the operating temperature is increased. Once graphite has been heated and outgassed in a vacuum furnace, it no longer can give off any carbonaceous gases that can be carburizing to the P/M parts during sintering.

13.3 REFRACTORY-METAL HEATING ELEMENTS:

(a) Molybdenum: Molybdenum elements often overlap or take up where silicon-carbide elements leave off in the high temperature range. There are two element designs used in sintering furnaces:

(1) In the small-size furnaces, the molybdenum elements are wound directly onto alumina muffles of round or "D" shape. Coatings of cement are then applied over the windings to hold them in place. The muffle is placed in a gas-tight welded shell and insulated with a high purity alumina refractory. The leads from the element must be sealed by means of packing glands. This design is chiefly employed on furnaces of the laboratory size having muffles from 3 in. to 6 in. (7.6 to 15.2 cm) in diameter or width. The construction is not suitable on larger furnaces because the refractory will sag at the high temperature and break the element.

(2) In the larger furnaces the design employed is identical to nickel-chromium design of rod overbend or ribbon overbend elements. The construction allows direct radiation of heat to the work. The elements are supported on the walls of the furnace by molybdenum pins or hanger bolts, or refractory supports. The furnace may be lined with additional insulation for the higher-temperature operation. If a dry-hydrogen atmosphere is to be maintained, the refractory must be a high-purity alumina brick to prevent deterioration in the purity of hydrogen atmosphere. Molybdenum and high-purity alumina refractory are expensive. Therefore, a molybdenum element high-temperature sintering furnace can be a very expensive item when compared to the same size nickel-chromium or silicon-carbide furnaces. On the other hand, elements properly applied and cared for do not deteriorate as do the others. Barring accidents, they may last many years and, therefore, can have the lowest maintenance cost of all.

Molybdenum has the characteristic of becoming coarse-grained and quite brittle after it is heated, more so at room temperature than at high temperature. Therefore, care must be taken not to thermally shock the element by too great a change in electrical input, or by charging a load of cold work directly into the furnace without a preheat. The resistance of molybdenum is much lower at room temperature than at elevated operating furnace temperatures, and therefore starting taps of lower voltage must be provided so that the element is not subjected to thermal shock or overloading. The transformer, however, is not as expensive as the silicon-carbide-element transformers which must have many voltage taps. Saturable-core reactor control is often preferred to on-off control to provide unattended, safe start-up for the molybdenum heating elements. Also, by increasing the voltage on the elements as the temperature and resistance increase, while holding constant current, the power input increases. This is a desirable feature, since more energy usually is needed at higher temperatures.

Molybdenum elements can be operated up to 3400° F (1890° C) element temperature in a hydrogen atmosphere, but consideration must be given to the use of proper refractories. The watt density loading for molybdenum can be as high as 50 to 60 watts per square inch (7.75 to 9.30 watts/cm^2). This allows considerable input into a small chamber which is required for high temperature operation, to overcome radiation losses. In designing a molybdenum element, the current density (amperes per square inch or sq cm of cross-section) must also be taken into the design calculations, as well as the watt density loading.

Molybdenum elements can be welded and spliced, but the welds are brittle and must be handled with great care. Warming up helps their ductility. Where possible, the element is usually made of one continuous piece of molybdenum wire. The lead-out terminals are made to have low resistance, such as by twisting several extra strands of molybdenum wire around the terminals so excessive temperature will not be developed in them.

Molybdenum furnaces are used for sintering stainless steels, refractory metals and cermets; carburizing tungsten to make tungsten carbide, and sintering tungsten carbide to make tool and cutting tips.

(b) Tungsten: Tungsten heating elements are used specifically in vacuum sintering furnaces for application above 3300° F (1833° C), the top limit for molybdenum heating elements. Tungsten is more expensive and more difficult to form than molybdenum and is used only for vacuum furnaces at temperature ranges from 3300° F to 4500° F (1833° to 2500° C) to sinter refractory metals and carbides.

14.0 EFFECTS OF FURNACE ATMOSPHERES ON HEATING ELEMENTS

It is very difficult to compare the life of heating elements in various furnace atmospheres in terms of expected hours of operation. The life of a resistance material depends not only on furnace atmosphere, but also on many other working conditions, i.e., watt density loading, cross-sectional area, temperature of operation, frequency of switching the current on and off, support of the resistor in the furnace, physical shape and design of the resistor, and how hard the furnace is worked.

The experienced furnace designer knows the shortcomings of a particular resistor to a specific furnace atmosphere, and designs more conservatively than he would for an atmosphere that has little effect on the resistor material. Thus, a comparison in hours for the same resistor in commercial furnaces operated at the same temperature would not give a true picture of the action of the different atmospheres. The conservative design would not be used for the more favorable atmospheres because of economic considerations. Therefore any data available on hours of operation would be misleading and very controversial in nature.

TABLE 14.1

MAXIMUM RECOMMENDED FURNACE OPERATING TEMPERATURES FOR TYPICAL HEATING ELEMENTS IN DIFFERENT FURNACE ATMOSPHERES(a)

Furnace Atmosphere Gases(b)	Air: Natural Gas Ratio	Dew Point °F Entering Furnace	Heating Element Materials (g). Temperature Indicated in °F					
			35Ni-20Cr-45 Fe	80Ni-20Cr	Moly	Tungsten	Silicon Carbide	Graphite(c)
1 Hydrogen								
A. By Electrolysis of Water								
1. Direct from Cells								
a. Unpurified—Saturated		+70 to +90	1800	2150 (d)	3200	3200	2350	NR
b. Purified		-80 or lower	1800	2050 (i)	3400 (i)	4500 (i)	CR-2100 (i)	4000
B. From Bottles								
1. Unpurified		-30 to +20	1800	2050 (d)	3200	3200	2350	NR
2. Purified		-80 or lower	1800	2050 (i)	3400 (i)	4500 (i)	CR-2100 (i)	4000
C. By Catalytic Conversion of Hydrocarbons								
1. Unpurified—Saturated		+70 to +90	1800	2150 (d)	3200	3200	2350	NR
2. Dried		-100 or lower	1800	2050 (i)	3400 (i)	4500 (i)	CR-2100 (i)	4000
D. From Liquid Hydrogen		-84 or lower	1800	2050 (i)	3400 (i)	4500 (i)	CR-2100 (i)	4000
2 Nitrogen—Bottled		-64 or lower	1800	2100	3000 (k)	3000 (k)	2400	5000
3 Hydrogen-Nitrogen Mixtures								
A. Dissociated Ammonia								
1. As Reacted—Dry		-60 to -40	1800	2050 (i)	3200 (i)	4500 (i)	CR-2100 (i)	4000
2. Moisture Added—Saturated		+70 to +100	1800	2150 (d)	3000 (k)	3000 (k)	2350	NR
B. Burned Dissociated Ammonia								
1. Rich—Saturated		+70 to +90	1800	2150 (d)	3000	3000	2350	NR
2. Lean—Saturated		+70 to +90	1800	2100	3000 (k)	3000 (k)	2350	NR
C. Direct Catalytic Conversion of Ammonia and Air								
1. Rich								
a. As Reacted—Saturated		+70 to +90	1800	2150 (d)	3000	3000	2350	NR
b. Cooled to 40°F		+40	1800	2100 (d)	3000	3000	2350	NR
c. Dried		-80 or lower	1800	2050 (i)	3200 (i)	4500 (i)	CR-2100 (i)	4000
2. Lean								
a. As Reacted—Saturated		+70 to +90	1800	2100	3000 (k)	3000 (k)	2350	NR
b. Cooled to 40°F		+40	1800	2100	3000 (k)	3000 (k)	2350	NR
c. Dried		-80 or lower	1800	2100 (i)	3200 (i),(k)	4100 (i),(k)	CR-2100 (i)	4000
4 Reformed Hydrocarbon Gases								
A. Exothermic Gas								
1. Rich								
a. As Reacted—Saturated	6:1	+70 to +90	1800	2150 (d)	NR	NR	2350	NR
b. Refrigerated	6:1	+40 to +60	1800	2100 (d)	2500	2500	2350	NR
2. Medium Rich—Saturated	6.75:1	+70 to +90	1800	2150 (d)	NR	NR	2350	NR
3. Lean—Saturated	10.25:1	+70 to +90	1800	2100 (d)	NR	NR	2350	NR
B. Purified Exothermic Gas								
1. Rich	6:1	-40 or lower	1800	2000	NR	NR	2350	4000
2. Medium Rich								
a. As Reacted	6.75:1	-40 or lower	1800	2000	NR	NR	2350	4000
b. Methane Added	6.75:1	-40 or lower	NR	CR (e)	NR	NR	2350 (h)	4000
3. Lean	10.25:1	-40 or lower	1800	2100	NR	NR	2350	4000
C. Endothermic Gas								
1. Rich—Dry	2.4:1	-10 to +10	CR-1800	2000	NR	NR	2350	NR
2. Rich—Fairly Dry								
a. As Reacted	2.6:1	+20 to +30	1800	2000	NR	NR	2350	NR
b. Methane Added	2.6:1	+20 to +30	NR	CR (e)	NR	NR	2350 (h)	NR
3. Medium Rich—Saturated	3.5:1	+70 to +90	1800	2150 (d)	NR	NR	2350	NR
4. Lean—Saturated	4.5:1	+70 to +90	1800	2150 (d)	NR	NR	2350	NR
5 Argon—Bottled		-73 or lower	1800	2100	3200 (k)	4100 (k)	2600	5000
6 Helium—Bottled		-73 or lower	1800	2100	3200 (k)	4100 (k)	2600	5000
1-6 Above Atmospheres Containing Impurities of:								
A. Sulphur							Temps. Same	
1. Up to 10 Grains/C cu. ft.			CR-1800	CR-2050	NR	NR	as Above (f)	NR
2. 10 to 25 Grains/C cu. ft.			Cr-1800	NR	NR	NR	Temps. Same	NR
B. Lead			NR	NR	NR	NR	as Above (f)	NR
C. Zinc								
1. From Brass, Openly Loaded			CR	CR	NR	NR	Temps. Same	NR
2. From Zinc Stearate			1800 (h)	2050 (h)	NR	NR	as Above (h)	NR
7 Vacuum (Below 150 Microns)			NR	1850 (j)	3300 (j)	4500 (j)	2100 (j)	4000 (j)
8 Air (g)								
A. Normal		+50 or lower	1700	2100	NR	NR	2600	NR
B. Wet		Above +50	1700	2100	NR	NR	CR	NR

Footnotes to Table 14.I

(a) The purpose of this information is to serve only as a selection guide. It is not to be used as an operating manual, or in any way to conflict with nameplate data, operating instructions, or bulletins supplied with or related to any furnace. Data pertain only to heating elements. Lower maximum temperatures often are necessary due to limitations of refractories and supports.

Temperatures listed for all materials, except graphite, can be taken as furnace temperatures, provided elements are designed with conservatively low watts release for good life at maximum operating temperature.

Data given are based on typical heating elements in general use for powder metallurgy as of November 1, 1976

(b) For approximate compositions and costs of atmosphere gases see Table 19.1

(c) Element temperatures (not ambient) commonly used in controlling graphite tube furnaces. In most cases life is measured in weeks, but is considered acceptable in order to achieve the high temperatures indicated.

(d) Due to "green rot" failure by wet H_2 and/or CO (but not when dry), life may be reduced substantially at temperatures below 1800° F. The 35 Ni-20 Cr Alloy is not subject to "green rot" and is preferred to 80 Ni-20 Cr below 1800° F.

(e) Special 80 Ni-20 Cr elements with ceramic protective coatings designed for low voltage (8-16 volts) operation can be used successfully.

(f) Use 2400° F maximum for argon and helium with sulphur impurities instead of 2600° F maximum listed above.

(g) Elements not listed above, useful primarily in air, are as follows:
- (1) 23Cr-4.5Al-2 Co-Bal. Fe—Good to 2100° F in air. Fair to 2100° F, if oxidized prior to admitting H_2 atmosphere.
- (2) 24Cr-6.0Al-2 Co-Bal. Fe—Good to 2400° F in air. Fair to 2400° F, if oxidized prior to admitting H_2 atmosphere.
- (3) 37Cr-7.5Al-Bal. Fe —Good to 2400° F in air. Fair to 2400° F, if oxidized prior to admitting H_2 atmosphere.
- (4) Platinum —Good to 2550° F in air.

(h) When heating elements and walls are exposed to deposits of carbon and zinc from volatilized lubricants, or to carbon from methane and propane, these deposits combine with, deteriorate, and short circuit heating elements, and overload control and power supplies. Such deposits should be burned out with air frequently at 1200 to 1400° F, to help element life and preserve the equipment. Carbon combined with heating elements or brickwork cannot be "burned out," so diligence is required to prevent or minimize its presence on and in them. This becomes increasingly important with increasing dryness of atmospheres. Moisture in the atmosphere reacts with the volatiles and helps to minimize or "clean up" such deposits. Dry atmospheres do not give this protection.

(i) Data based on assumption that precautions will be taken to assure that high-purity atmosphere will be maintained within furnace chamber surrounding heating elements. For example, for dry-hydrogen applications, it is assumed that the dew point of effluent gas from the heating chamber will be about −40° F or drier. The sample must be taken at a distance from the inlet so as to include contaminants.

(j) Vacuum furnace maximum temperatures are only conditionally recommended, and life may be short or long due to variable design and operating conditions including: Time-temperature-pressure cycles; reactions with various contaminating gases during degassing periods at different temperatures; compositions, volatilization and creep rates of element materials; progressive mechanical distortion of heating elements within the plastic deformation range; recrystallization under heat; and mechanical abuse during loading and unloading.

(k) Precautions are necessary to prevent oxygen contamination of atmosphere by infiltrating air, since free oxygen can cause damaging oxidation of moly or tungsten heating elements. Such protection can be had by using purging chambers, or by mixing a small amount of hydrogen with the atmosphere gas to react with oxygen to form H_2O and thereby protect elements from oxidizing. In the latter event, dryness of the furnace atmosphere to satisfy footnote (i) would be destroyed, and maximum temperature should be about 3000° F.

NR-Not recommended as one of the best element materials for this atmosphere.

CR-Conditionally recommended where shorter life is acceptable. For recommendations consult furnace manufacturer.

To help select the proper resistor material as a heating element in a known furnace atmosphere, Table 14.1 was prepared. The table is based on the recommended furnace operating temperatures for typical heating-element resistor materials for the specific atmospheres. This gives the engineer a choice of materials, depending upon the maximum operating temperature of each furnace. Resistor temperatures are always higher than furnace control temperatures, and the difference depends upon the design of watt density loading of the resistor. Therefore, when the furnace is to be operated close to the maximum temperature, the watt density loading must be lower and more conservative. If this rule is kept in mind, plus the rule that the cross-sectional area of the heating element should be increased for furnace atmospheres that show more attack than others, Table 14.1 will serve as guide to the proper selection of the resistor material for a specific application, and should lead to a workable design. To supplement Table 14.1, the following brief discussions of the action of various furnace atmospheres on heating-element resistor materials may be helpful.

14.1 OXIDIZING ATMOSPHERE— AIR:

With the exception of air sintering for aluminum P/M parts little if any sintering is done in air atmospheres. However, the atmosphere outside metal muffles or retorts is often air, so some comments regarding heating elements in air atmospheres may be helpful.

With the exception of molybdenum, tungsten and graphite, all other resistor materials commonly used have maximum life when operating in an oxidizing atmosphere of air. Air forms an impervious, oxide protective film to reduce further attack by oxidation. The protective oxide coating produced by air on the nickel-chromium series is so impervious that the base metal in the core of overheated elements may melt and run out at the overheated sections, leaving the oxide skin of the original shape of the heating element intact on the walls of the furnace. The inexperienced operator often claims the failure is due to a hollow defect in the resistor material, instead of a defect in the temperature control causing overheating.

Since the 80 nickel-20 chromium resistors have the greatest resistance to oxidation, and can be used to operate furnace temperatures up to 2100° F (1150° C), this is the alloy most commonly used in air atmospheres outside muffles and retorts.

The chromium-aluminum-iron alloy series, which may or may not contain a small percentage of cobalt, are used primarily in the temperature range of 2100° - 2400° F (1150° - 1315° C) in air.

For element temperatures up to 2900° F (1590° C) in air, the non-metallic, silicon-carbide type is commonly employed. The maximum life of the resistor is obtained when operating in an atmosphere of dry rather than moist air. This resistor is more commonly used than the chromium-aluminum-iron type.

Molybdenum-disilicide elements, in relatively small sintering furnaces, are used up to 2900° F (1590° C) in air. Because of the high cost of platinum, it is used only in very special cases, where silicon-carbide cannot be worked into the furnace design. Platinum is restricted to air, and cannot be used in reducing atmospheres.

14.2 REDUCING ATMOSPHERES:

Reducing atmospheres may be divided into four classes: (1) dry hydrogen or dissociated ammonia containing no oxidizing or carburizing constituents; (2) unpurified rich exothermic atmosphere having a low carbon potential or being decarburizing in nature; (3) purified rich exothermic atmosphere having medium carbon potential and sometimes enriched for high carbon potential; and dry endothermic or charcoal atmospheres having a high carbon potential; (4) dry endothermic atmosphere enriched with a hydrocarbon gas to produce a carburizing atmosphere, also enriched with ammonia and hydrocarbon to produce a carbonitriding atmosphere.

The term "reducing atmosphere" is in reference to the atmospheric reaction with iron and iron oxide.

With the exception of dry hydrogen or dissociated ammonia, all the above atmospheres are oxidizing to the nickel-chromium series. Even the hydrogen or dissociated ammonia atmospheres will oxidize chromium, unless the gas is extremely dry. The type of oxide produced by the so-called reducing atmosphere is entirely different than that produced by air. The oxide produced by air is a green to black, impervious type that retards further oxidation of the metal below its surface. The oxide produced on high-nickel-chromium elements by wet reducing furnace atmospheres is green and not impervious, and the atmosphere keeps attacking the base metal. This type of attack has generally been referred to as "green rot". It occurs in some alloys such as 80Ni-20Cr in wet hydrogen, rich exothermic and wet endothermic gases in a limited temperature range of 1650° to 1850° F (900° - 1010° C) element temperature. The 35Ni-20Cr series with at least 1.25% silicon, however, is quite resistant to green rot attack. Accordingly, 35Ni-20Cr is recommended in wet hydrogen-bearing gases up to 1800° F (980° C) furnace temperature, and 80Ni-20Cr above 1800° F (980° C).

However, using a columbium stabilized 80Ni-20Cr alloy (1.25 Cb) will, for most practical applications and atmosphere conditions, eliminate or greatly minimize effects of "green rot".

Dry hydrogen and dissociated ammonia have the least effect of the listed reducing atmospheres on the nickel-chromium group. At elevated temperatures above 2000° F (1090° C), the resistor may have better life in dry hydrogen than in air, because the oxidation rate in air becomes more rapid at elevated temperatures.

Above 2000° F (1090° C) furnace temperature the unpurified rich exothermic atmosphere has less harmful effect than the higher carbon potential purified rich exothermic, dry endothermic, or charcoal atmospheres on 80Ni-20Cr. The higher carbon potential atmospheres carburize the nickel-chromium alloys, especially at the higher temperatures. Chromium is a strong carbide-former, and may pick up enough carbon to lower the melting point of the alloy and cause localized melting and fusion in the heating element. For this reason, it is safer to limit the operating temperature of 80Ni-20Cr to 2000° F (1090° C) maximum, when operating in reducing atmospheres of high carbon potential, unless the voltage can be reduced to lower the element temperature. Below 1900° F (1040° C) the 35Ni-20Cr alloy is best, despite the carburizing effect which does not necessarily mean that the effectiveness of the heating element is impaired.

Carburizing atmospheres produced by adding a hydrocarbon to purified rich exothermic gas, and dry endothermic gas, or by adding ammonia as well as a hydrocarbon to produce a carbonitriding atmosphere are not recommended for use on unprotected heating elements made of the nickel-chromium alloys. The short life of the resistor in this type of atmosphere is due to the carburizing effect, and also to shorting and melting out of the leads and terminals because of carbon depositing on and becoming impregnated in the refractory lining. A special nickel-chromium heating element has been developed for operating in a carburizing atmosphere by protecting the alloy with a ceramic, high temperature enamel to resist carburization. The heating element operates on low voltage (8 - 10 volts), to prevent arcing at the terminals in the carbon-impregnated brickwork.

The chromium-aluminum-iron resistor alloys cannot be used successfully in any reducing atmosphere. Some success may be had in dry hydrogen if the alloy is previously oxidized in air, but the results are questionable. The chromium-aluminum-iron series depends upon the impervious oxide coating produced by air for its protection by high temperature oxidation, and therefore reducing or carburizing atmospheres are not recommended.

Molybdenum has excellent life up to 3400° F (1890° C) in dry hydrogen, dissociated ammonia, low-hydrogen mixtures of nitrogen, argon, or helium, or in rich exothermic gas of 60° F (16° C) dew point or less with 2500° F (1370° C) maximum operating temperature. In those gases, where hydrogen is absent or in minute quantity, a little hydrogen addition is considered desirable to help avoid oxidation of molybdenum by infiltrating air. Poor life results because of carburization when molybdenum is operated in a high carbon potential atmosphere. Air or high moisture will oxidize the elements at elevated temperatures. Molybdic oxide (MoO_3) is volatile when formed at high temperatures, and thus no protective oxide coating adheres to the molybdenum. If the proper atmosphere is not applied, the elements will volatilize and fail. When started up, the furnace should be purged with non-flammable atmosphere and then with hydrogen before turning on the power. In shutting down, protective atmosphere should be maintained to about 700° F (370° C) to avoid harmful oxidation of the molybdenum by air.

Graphite resistors have been used for high-temperature furnace applications in atmospheres free of oxidizing constituents of oxygen, carbon dioxide and water.

Silicon-carbide resistors have good life in some reducing atmospheres. However, the maximum operating temperature is reduced from that when operating in air. Silicon-carbide elements also have the advantage that they can be operated in high carbon potential atmospheres that would easily carburize and destroy nickel-chromium elements at high sintering temperatures. The carbon deposited on the silicon-carbide resistors by stearate carry-over will eventually cause the resistors to draw more current, but this can be seen on the ammeter of the transformer, and the voltage to the resistors can easily be dropped by setting the transformer to a lower voltage tap until time permits the carbon to be burned out of the furnace.

14.3 ATMOSPHERES CONTAINING CONTAMINANTS OF SULPHUR, LEAD OR ZINC:

Sulphur, if present, will appear as hydrogen sulphide in reducing atmospheres and as sulphur dioxide in oxidizing atmospheres. The source of sulphur contamination is usually due to one of the following: (1) high sulphur content of fuel gas used for producing the protective atmosphere, which generally can be eliminated; (2) high-sulphur-bearing refractories or cement; or (3) the material being processed in the furnace.

Sulphur is very poisonous to the nickel-chromium group of resistors in reducing atmospheres, and causes pitting and blistering of the resistor alloy in oxidizing atmospheres. The higher the nickel content of the resistor, the greater the attack will be. Therefore, if sulphur contamination is present and cannot be eliminated, the 35Ni-20Cr alloy will give the best performance.

Lead and zinc contamination results from the work being processed in the furnace. This is common in sintering furnaces for processing powder metallurgy parts. In the presence of a reducing atmosphere, lead vapor will leave a lead-bronze mix and contact the heating elements. Lead metallic vapors are even more poisonous to the nickel-chromium group than sulphur, and can damage the resistor in a matter of hours, under certain conditions of concentration and temperature. The higher-nickel alloys are more affected than the lower nickel alloys.

Zinc contamination results from zinc stearate used as a lubricant when pressing P/M compacts. The zinc stearate volatilizes out of the compacts when heated. Zinc vapors alloy with nickel-chromium heating elements, resulting in poor life. To eliminate the zinc vapors at the higher sintering temperatures, the common practice is to use a burn-off chamber operated at about 1200° F (650° C), as previously discussed. If this cannot be done, or if zinc is present in the metal being processed, then the non-metallic silicon-carbide resistors should be used. Silicon-carbide resistors are not affected by sulphur, lead or zinc contamination, and therefore are commonly used at low temperatures as well as high temperatures, when contamination is anticipated.

14.4 VACUUM ATMOSPHERE:

For applications involving vacuum 80Ni-20Cr has been used successfully up to temperatures of 1850° F (1010° C) when the element is directly exposed to a low pressure. In furnaces where the elements are external to the low pressure chamber, or where the elements are exposed to a pressure above 100 microns, 80Ni-20Cr has been used successfully up to 2100° F (1150° C).

The 80Ni-20Cr alloy is not successful above 1850° F (1010° C) in low pressures, because the vapor pressure of chromium is high enough at elevated temperatures to remove the chromium from the resistors, resulting in poor life, contamination of the furnace and material being processed, and loss of vacuum. Because of this, the watt density loading at low pressures must be kept low and very conservative, especially at the higher operating temperatures. For furnace temperatures from 1850° - 3300° F (1010° - 1830° C), molybdenum is the best choice and gives satisfactory life. Tungsten and tantalum have been used successfully for furnace temperatures up to 4500° F (2500° C), silicon carbide up to 2100° F (1150° C) at 3 microns pressure. Solid graphite, rod or plate elements are being used up to about 4000° F (2200° C) element temperature.

Graphite cloth elements for both heat treating and sintering of P/M parts are now available and are being used over a wide temperature range; 1600° F to 3100° F (871° C to 1650° C).

When a vacuum sintering furnace which is constructed of fibrous graphite felt and cloth is heated, a chemical reaction occurs between the graphite and the residual gases. This reaction alters the relationship between the partial pressure of the oxidizers and the total pressure of the system. Although the total pressure may remain stable, that is in quantity, the quality of the atmosphere improves due to the thermo-chemical reaction.

Specifically, the threshold of oxidation of graphite in air, water vapor, or carbon dioxide is:

O_2 = 750° F (399° C)
H_2O = 1350° F (732° C)
CO_2 = 1850° F (1010° C)

Therefore, when a furnace containing a residual atmosphere of these combined gases is heated above the oxidation thresholds listed above, the partial pressure of these residual gases is reduced.

In a furnace constructed of materials such as molybdenum and tungsten, which when heated will not react with the residual gases in the system, the reduction of the partial pressure of the oxidizers within that particular heating chamber can only be effected by a further reduction of the total presure in the system.

Operating pressures of mm Hg (10^{-4} Torr) (diffusion pump range) are customary for general heat treating in a furnace with metallic radiation shields. In a pure graphite lined furnace, pressures of 50-100 microns, which can be easily be attained with only a mechanical pump, may be used.

15.0 EFFECTS OF FURNACE ATMOSPHERES ON REFRACTORIES

The selection of the proper refractory material is an important consideration in sintering furnaces that have the refractory lining exposed to the furnace atmosphere. Various types of furnace atmospheres have different effects on refractories, and conversely, the refractories have an effect on the furnace atmosphere. If the proper refractory is not selected, very high maintenance may result. If extremely low dew points are required, as in the case of sintering stainless steel, the selection of a refractory that will not react with dry hydrogen to form water vapor is very important if the furnace is of the non-muffle type.

15.1 EFFECT OF CARBON MONOXIDE ON REFRACTORIES:

Both endothermic and exothermic reducing atmospheres contain carbon monoxide. Carbon monoxide is stable at elevated sintering temperatures, but under some conditions breaks down into carbon and carbon dioxide in the temperature range between 800° - 1200° F (427° - 649° C). When an atmosphere gas containing carbon monoxide is free of carbon dioxide and low in water it is unstable and needs only the promoting effect afforded by a catalytic surface to cause a rapid reaction to take place.

FIGURE 15A.

Carbon globules in a light-weight insulating brick, due to reversible reaction of carbon monoxide promoted by iron oxide in the brick

The equilibrium of the reaction may be represented in the following equation: $2\ CO = C + CO_2$. Iron oxide and certain other oxides such as chromium and nickel act as a catalyst for this reaction. Therefore, in selecting refractories that are exposed to carbon monoxide, it is important to select as low an iron-oxide content as practical.

The iron-oxide particles act as nuclei to start the reversible reaction. Carbon then begins to deposit in globules, as shown in figure 15A. The globules then keep growing until the carbon ruptures and disintegrates the refractory brick. This reaction is, of course, more critical in the zones where the temperature may be between 800° F and 1200° F (427° - 649° C), such as burn-off and purge zones and insulated cooling sections. However, the refractory in the high-temperature sintering section is also subjected to this reaction because at some point below the hot face of the refractory there will be a zone between 800° F and 1200° F (427° - 649° C). This effect is shown in a light-weight insulating brick in figure 15B.

FIGURE 15B.

Carbon core in light-weight insulating brick caused by reversible reaction of carbon monoxide in temperature range of 800°F to 1200°F (427° - 649°C)

The same reaction, of course, also occurs in hard fire brick as well as light-weight insulating fire brick. In hard fire brick another reaction, besides the carbon deposition, seems to occur which breaks down the bond and the refractory disintegrates into dust. This is shown in figure 15C.

FIGURE 15C.
Carbon globules and disintegration of bond in hard fire brick caused by reversible reaction of carbon monoxide promoted by iron oxide in the refractory

Some furnace manufacturers have done considerable research on this subject, and have set up quality control on the refractory used in atmosphere furnaces. In extremely critical zones, such as "burn-off" and/or pre-heat chambers, and insulated cooling chambers, extra precaution should also be taken to use high-temperature refractory, 2800° F or 3000° F (1538° C or 1649° C), instead of the less-expensive lower-temperature grades, 1500° - 2600° F (816° - 1427° C). The reason for this is that any residual iron oxide in the brick would combine with the silica, alumina and clays at the high firing temperature the refractory was subjected to in manufacture, and thus, the free iron oxide will not be present to act as a catalyst. This, of course, leads to a more expensive furnace, but in the long run will pay many times over the initial cost in the savings from low furnace maintenance. The same rule applies to hard fire brick. Only super-duty fire brick should be used in furnaces containing a carbon monoxide atmosphere, regardless of the temperature of operation.

Deposition of carbon is harmful to the refractory and to heating elements mounted on the refractory face. As shown in figure 15B, the elements may short out between turns or between the lead-in terminals. Since carbon is an excellent conductor, the voltage should be kept as low as practical on controlled atmosphere furnaces.

Even with the proper refractory, the furnace must be operated and maintained properly to assure good performance and low maintenance. When carbon-monoxide type atmospheres are used, the furnace must be burnt out periodically to remove any carbon that may have deposited on the brick work or heating elements. On continuous furnaces, this should be done at least once a week, and more often if possible. This is especially important in furnaces where there may be a danger of carry-over of the lubricants. If excessive carbon builds up in the front of the sintering zone, steps should be taken to either (a) reduce the amount of lubricant used in the compact, (b) raise the temperature of the burn-off chamber, (c) increase the length of the burn-off chamber or add a separate burn-off chamber to the front of the furnace, or (d) improve the counterflow of atmosphere gas.

15.2 EFFECT OF HYDROGEN ON REFRACTORIES:

Under some furnace conditions, hydrogen will reduce the oxides of iron, silicon, and various other metals in most refractories except high-purity alumina. However, pure-alumina refractories can be reduced in a hydrogen atmosphere having an extremely low dew point, such as -140° F (-96° C), and a furnace temperature of 2400° F (1316° C). The higher the temperature, and the lower the dew point, the more severe the reaction and the greater the number of oxides reduced. This type of reaction will cause the bond to disintegrate, and the refractory will crumble. Where the refractory is exposed to high-hydrogen dry atmospheres in sintering furnaces, extreme care must be taken to choose the proper grade of refractory to withstand the temperature of operation.

15.3 EFFECT OF THE REDUCING REACTION ON DEW POINT:

When oxides are reduced, water vapor will be formed and the dew point of the atmosphere in the furnace will rise, regardless of the entering dew point. In molybdenum furnaces, where it is impossible to use alloy muffles because of the high operating temperature, the choice of the refractory for the hydrogen atmosphere is extremely important when low dew points are required. The refractory must be of the highest alumina available. Any silicon oxide or soda (NaO) in the refractory may be reduced to silicon or sodium and volatilized, then reoxidized in the case of silicon, and form sodium hydroxide in the case of sodium. Both will condense in the cooler parts of the furnace in the form of a white powder or glass wool. Care must be taken that silica and soda be excluded from the binder of the alumina refractory, as well as the cement used for laying up the refractory.

15.4 EFFECT OF ATMOSPHERE ON REFRACTORY HEAT LOSS:

Hydrogen has approximately five times the thermal conductivity of air. Therefore, expect higher heat losses for an equivalent thickness of insulation when hydrogen-rich atmospheres are used. When designing a furnace of tight-shell-construction, this must be considered in calculating heat losses. Refractory and insulation manufacturers' data usually give the heat loss in terms of an air atmosphere. When high-hydrogen reducing atmospheres are employed, a greater thickness of insulation must be added to the furnace wall, or the shell will run hotter with greater heat loss.

The relative thermal conductivities for the various reducing atmospheres in reference to air are: hydrogen 5 times as great, dissociated ammonia 3.75 times as great, endothermic 2 times as great, and rich exothermic only slightly greater. Therefore, it is necessary to base the heat-loss calculations on the particular atmosphere to be used. The refractory manufacturer should be consulted on heat loss data for a specific atmosphere.

16.0 CONVEYING COMPACTS THROUGH FURNACES

There are four methods in common use to convey compacts through furnaces: the mesh-belt conveyor, the roller hearth, the pusher type and the walking beam. Each type has its specific application, depending on the production rate, the weight and size of the parts to be sintered, the material to be sintered, and the operating temperature.

16.1 MESH-BELT CONVEYOR FURNACE:

One of the most commonly used sintering furnaces for continuous production of small, light parts is the mesh-belt conveyor furnace. As shown in figures 16A and 16B, the furnace consists of a charge table for loading the parts on the belt, a purge and burn-off chamber to distill off the lubricants, a high-temperature chamber, an insulated cooling chamber, a water-cooled chamber, and a discharge table. The parts are loaded on a continuously driven alloy mesh belt at the front of the furnace, and are discharged at the rear. A variable-speed drive allows flexibility to meet the heating requirements of the work to be treated. The alloy belt limits the furnace operation to about 2100° F (1150° C). Also, the mesh belt limits the furnace length because the stretch of the belt increases as the total load in the furnace increases. At temperatures of 2000° - 2100° F (1093° - 1150° C) the belt loadings are light to medium weight to favor belt life. At lower temperatures heavier loadings are permissible.

The mesh-belt type of furnace handles low to medium-high production rates, depending upon loading density, heating time, soaking time, etc. Doors on a mesh-belt furnace usually are left open during operation. In selecting the atmosphere generating equipment, ample capacity must be available.

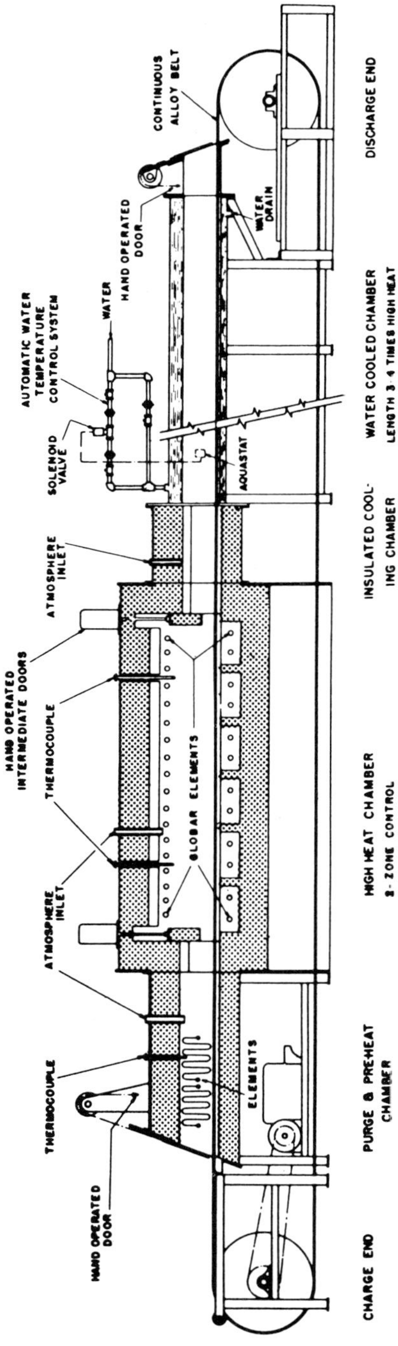

FIGURE 16A.

Longitudinal-section of a straight-through mesh-belt sintering furnace showing principles of construction and means of conveying work through the furnace

FIGURE 16B.
Installation of a mesh-belt continuous furnace for sintering iron P/M parts

16.2 HUMP-BACK TYPE FURNACE:

The hump-back mesh-belt conveyor furnace is a special type of continuous conveyor furnace used where high atmosphere purity is required.

A long gas-tight entry inclined purge chamber is used to carry the belt and work from the charge area up to the furnace hot zone, which is at an elevated level (figs. 16C and 16D). Following the hot zone, the cooling section is then inclined downward to a discharge point. Most large hump-back furnaces have a booster drive in the purge section to assist in carrying the work and belt up the incline. This reduces the stress on the mesh belt.

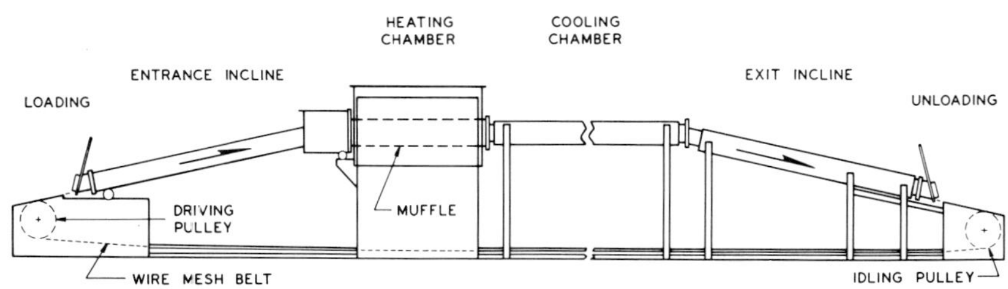

FIGURE 16C.
Typical arrangement of hump-back type mesh-belt furnace; omitted are details of belt drive and takeup mechanism and other features

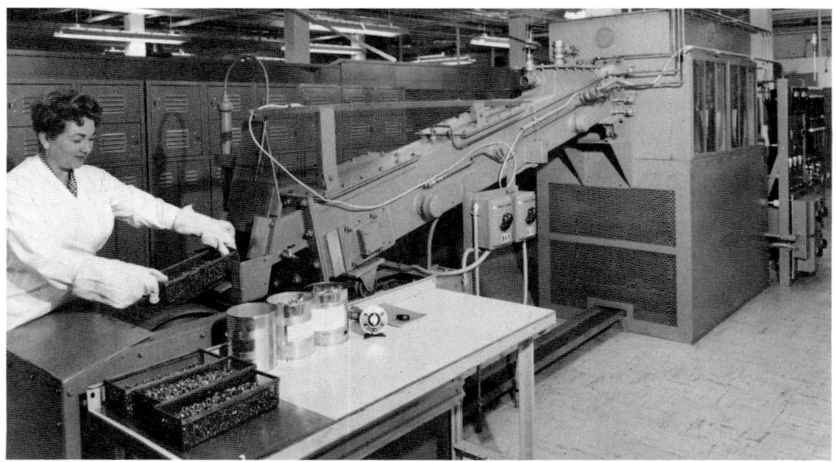

FIGURE 16D.
Hump-back mesh-belt furnace with alloy muffle, showing entrance incline and heating chamber

The hump-back furnace is particularly good where light atmospheres, such as dissociated ammonia or hydrogen are used. These gases tend to be confined in the furnace because of their natural tendency to rise. Because of this characteristic, this type of furnace generally operates with a lower gas consumption than a standard straight-through conveyor furnace. For the same reason, it generally is easier to establish a lower dew point and better atmosphere purity in a hump-back furnace.

The two major sintering applications which use this type of furnace are for stainless steel and aluminum. Each of these applications requires a very low dew point.

When a hump-back furnace is used, the burn-off of the compacts often is done in a physically separate furnace to reduce atmosphere contamination. The two furnaces may be linked together with a suitable conveyor to avoid separate handling of the work between the furnaces. An alternate arrangement is to burn off in the entrance inclined chamber.

Like the standard mesh-belt type of furnace, the hump-back is limited to a top temperature of about 2100° F (1150° C) maximum, and belt stress limits the length of furnace.

16.3 ROLLER-HEARTH CONTINUOUS FURNACE:

In the roller-hearth type continuous sintering furnace, trayloads of parts are conveyed by riding on driven rolls. The charge and discharge doors are automatically opened and closed by air or motor, and are interlocked with the

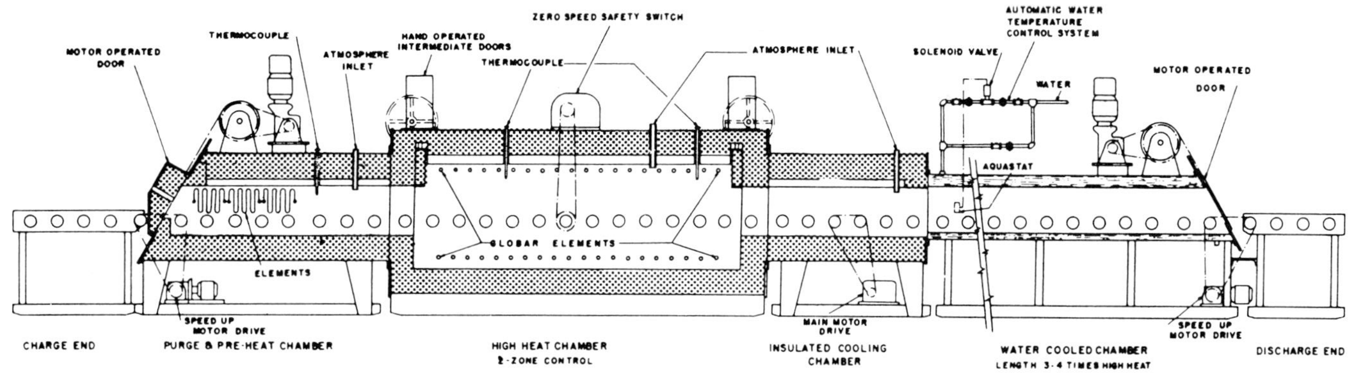

FIGURE 16E.

Longitudinal-section of a roller-hearth, continuous furnace showing principles of construction and means of conveying trayloads of P/M parts through the furnace

charging and discharging mechanisms. A longitudinal section of a roller-hearth furnace, outlining its components, is shown in figure 16E. The grade of alloy used in the rolls in the furnace limits its operation to 2100° - 2300° F (1150° - 1260° C). Each roll is driven, and depending on roll spacing, is capable of holding a load substantially greater than an equivalent length of mesh-belt conveyor.

The roller-hearth furnace is built in various widths and in any length to meet medium to high production requirements. The end doors on a roller-hearth furnace are opened only when a tray of work is charged or discharged, in order to economize on the amount of atmosphere gas required, and to minimize the heat losses for large openings. A typical roller-hearth furnace is shown in figure 16F.

FIGURE 16F.

Roller-hearth continuous furnace for sintering carbon-steel P/M parts in endothermic gas atmosphere

16.4 PUSHER-TYPE FURNACE:

The pusher-type sintering furnace is suited for sintering metal parts that load too heavily per lineal foot for the mesh-belt type, and where the production rate does not warrant the roller-hearth type furnace; or where the temperature for sintering is too high for either the mesh-belt or roller-hearth furnaces. Mechanical or hydraulic pusher furnaces are available for high-output capacities. Two types of pushing mechanisms are generally employed: the

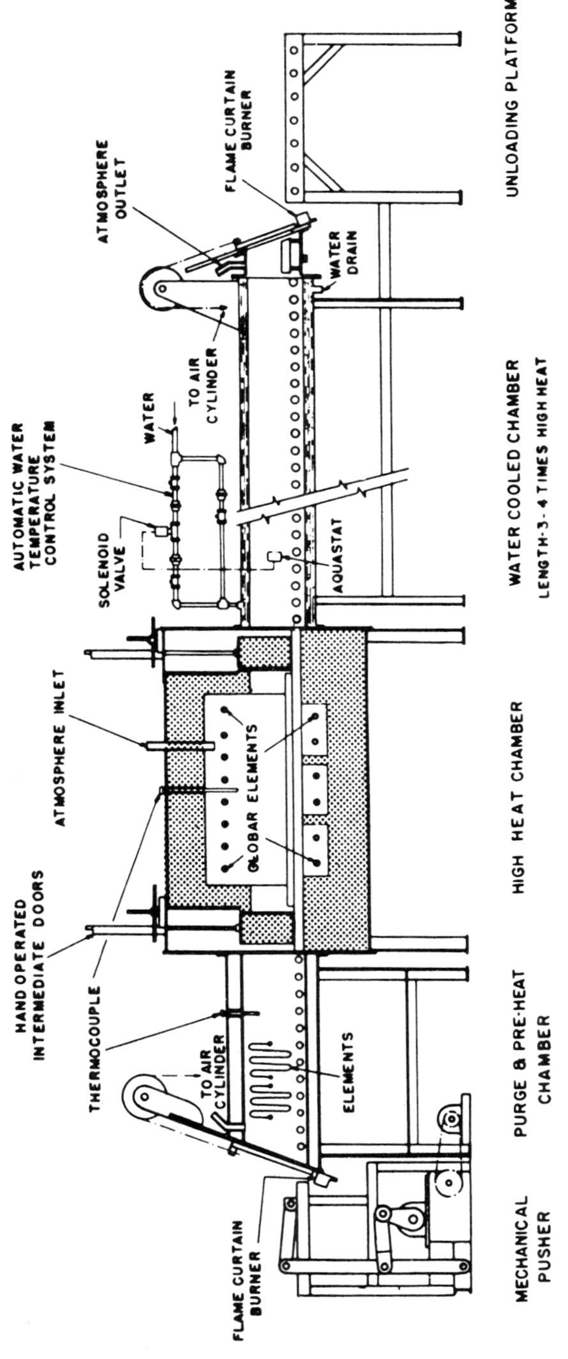

FIGURE 16G.
Longitudinal-section of a mechanical pusher, continuous furnace

intermittent pusher type, or the continuous stoker pusher type. The intermittent pusher mechanism normally is applied to bronze, brass, and iron P/M parts because of its low cost. The continuous stoker type is applied to furnaces for sintering carbides, stainless steel, and other applications wherein a gradual stepless temperature profile during heating and cooling is important. A typical pusher furnace is shown in figures 16G and 16H for sintering iron compacts. A full automatic pusher type, molybdenum heating element furnace is shown in figure 16I.

FIGURE 16H.
Installation of a mechanical pusher furnace with intermittent pusher, for sintering iron P/M parts in an endothermic atmosphere

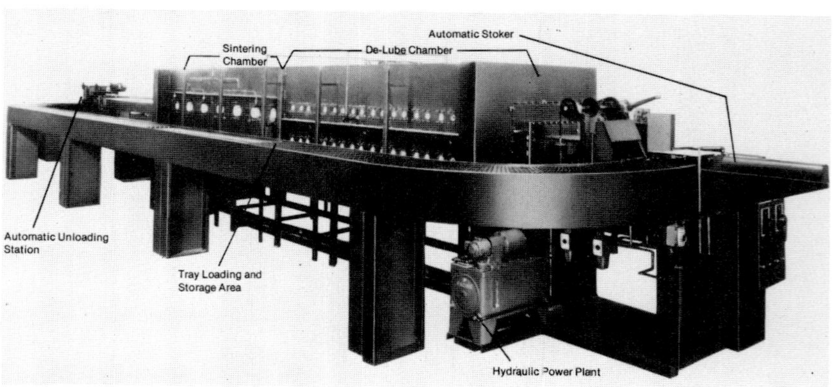

FIGURE 16I.
Automated pusher furnace with return conveyor

16.5 WALKING-BEAM FURNACE:

The walking-beam sintering furnace is another continuous type that can be used for the production of P/M parts (fig. 16J). Its advantages are:

(a) the amount of production work weight that can be safely conveyed through the furnace is practically unlimited.

(b) the maximum continuous operating temperature of the furnace is limited only by the refractory materials used to line the heating chamber, and the compatibility of the sintering atmosphere being used with the heating elements employed.

The walking-beam furnace is suitable for sintering P/M parts where the required production necessitates too high a load per linear foot of belt, and/or the process temperature is too high for a conveyor belt or roller hearth.

The operating temperatures may be as high as 3000° F (1650° C), depending on the compatibility of the atmosphere and the heating elements used.

The walking-beam is basically a mid-section of the floor throughout the length of the entire furnace, which is raised approximately ½ in. (1.27 cm) above the hearth level, pushed forward a predetermined distance, about 1 in. to 2 in. (2.54 to 5.08 cm), lowered ½ in. (1.27 cm) below the hearth, and pulled back to the starting position (fig. 16J). The up-and-down motion of the walking-beam is accomplished by means of a hydraulic cylinder (A) operated by an adjustable cycle timer and activated by a four-way valve (double solenoid operated). The cylinder (A) pushes and pulls a lever mechanism which, in turn, raises and lowers the conveyor. A second cylinder (B), which is also activated by the four-way valve (through the same cycle timer as the previous cylinder), provides the forward and reverse motion. The timed combination of the cylinders produces a rectangular motion, conveying the work at the required speed.

Another variation is the dual beam, very useful in furnaces wider than 18'' (figure 16K). The basic function of the dual beam is to provide center support for the boat during the lowering and retraction of the beam. Such support shortens the span and allows for lighter boats. Example: a 24'' boat resting on 2'' shelves would be spanning a distance of 20'' with a single beam. The maximum span with a dual beam is only 6 ¼''. The weight of the boat is thus reduced by at least a factor of 2.

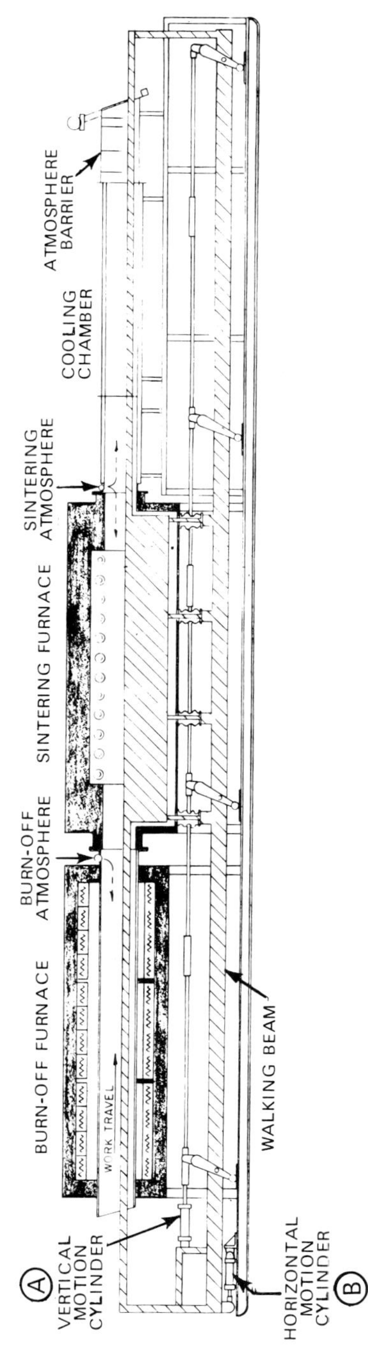

FIGURE 16J.
Longitudinal section of typical walking beam furnace

FIGURE 16K.
24" dual walking beam furnace for P/M compacts. Capacity 1,000 lbs. per hour. Maximum temperature rating 2400°F (1350°C)

16.6 BATCH-TYPE FURNACE:

Where production does not warrant continuous operation, or for experimental work, the pusher mechanism on mechanical pusher furnaces can be eliminated, and the work put through the furnace one tray at a time, by hand. Muffle or non-muffle batch furnaces with gas-tight interlocks and/or doors for charging and discharging work are also used for critical materials such as stainless steels or Alnico, where the dew point of the hydrogen atmosphere must be kept extremely low. A small hand-pusher muffle batch furnace with low-dew-point hydrogen or dissociated-ammonia atmospheres for sintering yellow brass is shown in figure 16L.

FIGURE 16L.
Installation of a full-muffle, hand-pusher furnace, for sintering small P/M parts

A box-type sintering furnace with burn-off chamber equipped with a muffle, and high-heat chamber without a muffle but with baffle doors at both ends, is shown in figure 16M.

Bell, elevator, or pit type furnaces also can be used as batch furnaces for sintering. The bell type furnace, figure 16N, is used widely for sintering P/M friction materials, such as elements for clutches, brakes, or other devices. Optional pressurizing mechanisms (not shown) are used for applying heavy loads to stacks of work pieces. For example, a pneumatically or hydraulically operated ram extending up through the center of the load; or a pneumatically operated diaphragm on top of the furnace, bearing down on top of the retort and load. Typical equipment consists of one or more stationary load-supporting bases with removable sealed retorts to cover the loads and retain protective atmospheres around them throughout the entire heating and cooling cycles, a portable heating bell and a standby base for it, a travelling hoist, and an optional cooling bell (not shown).

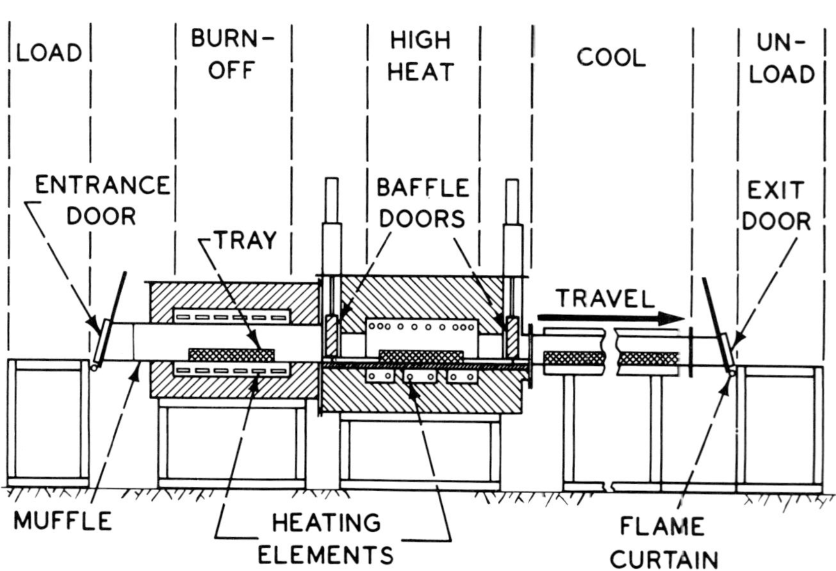

FIGURE 16M.
Box-type sintering furnace with burnoff chamber

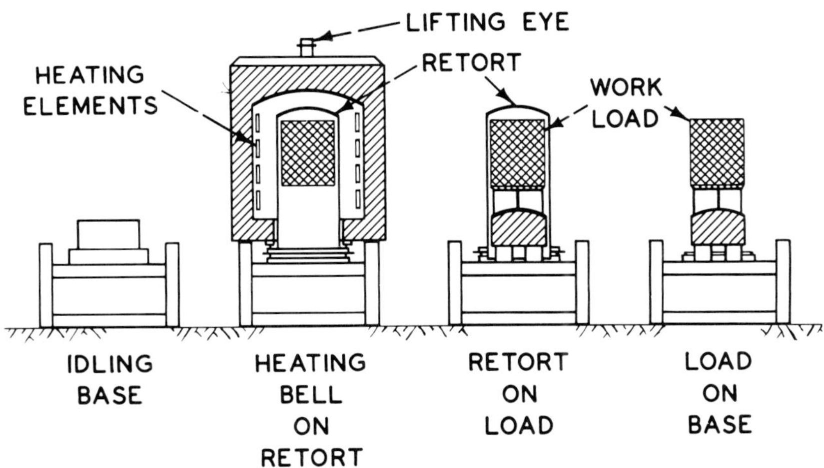

FIGURE 16N.

Schematic arrangement of bell-type sintering furnace showing stationary bases, retorts and heating bell

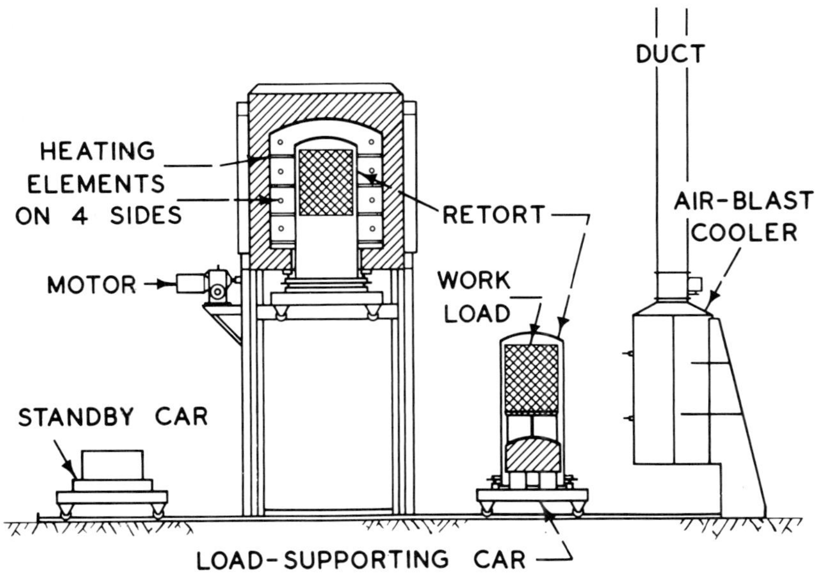

FIGURE 16O.

Sketch of typical elevator-type sintering furnace showing movable cars and stationary heating and cooling chamber

The elevator type furnace, figure 16O, is useful for sintering heavy or bulky loads, such as pressed-powder billets, stacked trayloads of compacts (fig. 16P) etc., with work temperatures up to about 2250° F (1230° C). Like the bell type, it is good for applications requiring protective atmospheres of exceptionally high purity, and is flexible for various time-temperature cycles in different types of loads. It has a fixed-position elevated heating chamber with open bottom, a mechanism for raising and lowering load-supporting cars in and out of the furnace, a standby car to plug the furnace opening during idling periods, and optional cooling chambers for hot loads from the furnace. Flexible hoses carry atmosphere gas and cooling water to and from the cars.

Vacuum furnaces employing the bell construction are used for sintering refractory metals. The metal powder is usually pre-sintered in a hydrogen-atmosphere furnace into the form of an electrode. The electrode is then placed in the vacuum bell furnace, and current is applied to the ends of the electrode to obtain the high temperature required for complete sintering.

Pre-sintered carbide tool tips are being production-sintered in vacuum furnaces inductively heated. Vacuum furnaces are being applied to other sintered products.

FIGURE 16P.

Elevator-type furnace for sintering stacked trayloads of stainless steel compacts

17.0 FURNACES FOR HEAT TREATMENTS OTHER THAN SINTERING

17.1 CARBURIZING AND CARBONITRIDING:

Iron and steel P/M parts can be heat treated after sintering, similar to parts made of wrought metals. The strength and hardness of medium and high-carbon steel parts can be increased by heating to 1500° - 1650° F (816° - 900° C) (depending on the carbon content) and quenching in oil. Low-carbon steel parts can be case-hardened by carbonitriding or carburizing, followed by an oil quench. Atmosphere methods for case-hardening are almost universally applied, because salt has a tendency to be absorbed into the pores of the parts, lowering their corrosion resistance. The best results of case-hardening are obtained on P/M parts having a density of 7.4 g/cc or higher. With a density of 7.4 g/cc, a well-defined martensitic case similar to that produced on wrought steel can be obtained.

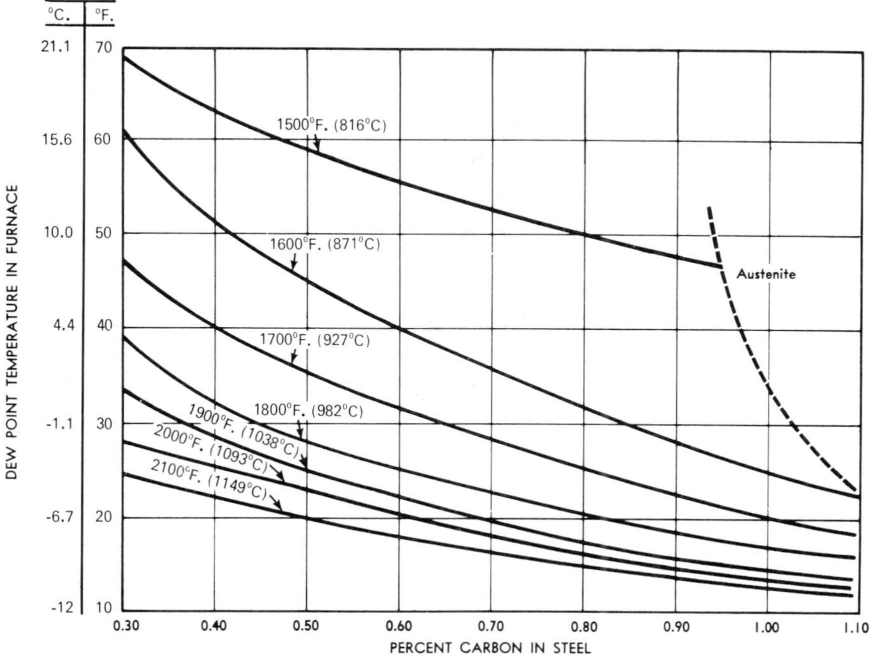

FIGURE 17A.

Equilibrium relationships between dew point and percent carbon in plain carbon steels at various furnace temperatures, when using an eńdothermic gas atmosphere. Curves were developed by direct measurment of dew point, temperature and percent carbon for plain carbon steels.

The atmosphere used for case-hardening is composed of the endothermic type previously described, adjusted to a high-carbon potential of 20° F (-6.7° C) dew point, with 5 to 10% natural gas added to increase the carburizing potential. If it is desired to obtain a case similar to that produced by cyanide, about 5% natural gas with 10% ammonia is added to the endothermic carrier gas. This treatment is referred to as carbonitriding. If the P/M part is composed of medium- or high-carbon powders, the dew point of the endothermic atmosphere is adjusted to the carbon content of the part, as shown in the equilibrium diagram, figure 17A.

The batch type, vertical gas fired radiant tube furnace shown in figure 17B is ideal for neutral hardening, carbonitriding, or straight gas carburizing. However, this furnace can also be furnished electrically heated, with a low voltage resistor developed for carburizing and carbonitriding atmospheres. This furnace can be supplied either completely automatic, or hand operated. The work is charged into the purge and quench vestibule by hand. In the automatic furnace, the work is carried into the furnace by a mechanical pusher. At the end of the heating cycle, the work is discharged automatically from the furnace onto the quench elevator, and then quenched. After quenching for the proper time, as set on the automatic timers, the work is raised automatically, and a signal notifies the operator to pull the tray out of the purge chamber. While one tray of work is in the quench tank, another is charged automatically into the furnace. Since the work does not contact air until completely heated and quenched, the parts stay bright and free from scale throughout the entire cycle.

A typical furnace for bright hardening or carburizing P/M parts is shown in figure 17C. Where production does not warrant automatic operation, or where it is desired to keep the initial cost low, the furnace can be operated manually. The above furnace has the capacity of neutral hardening up to 400 net pounds per hour, or light case carburizing or carbonitriding up to 300 pounds per hour.

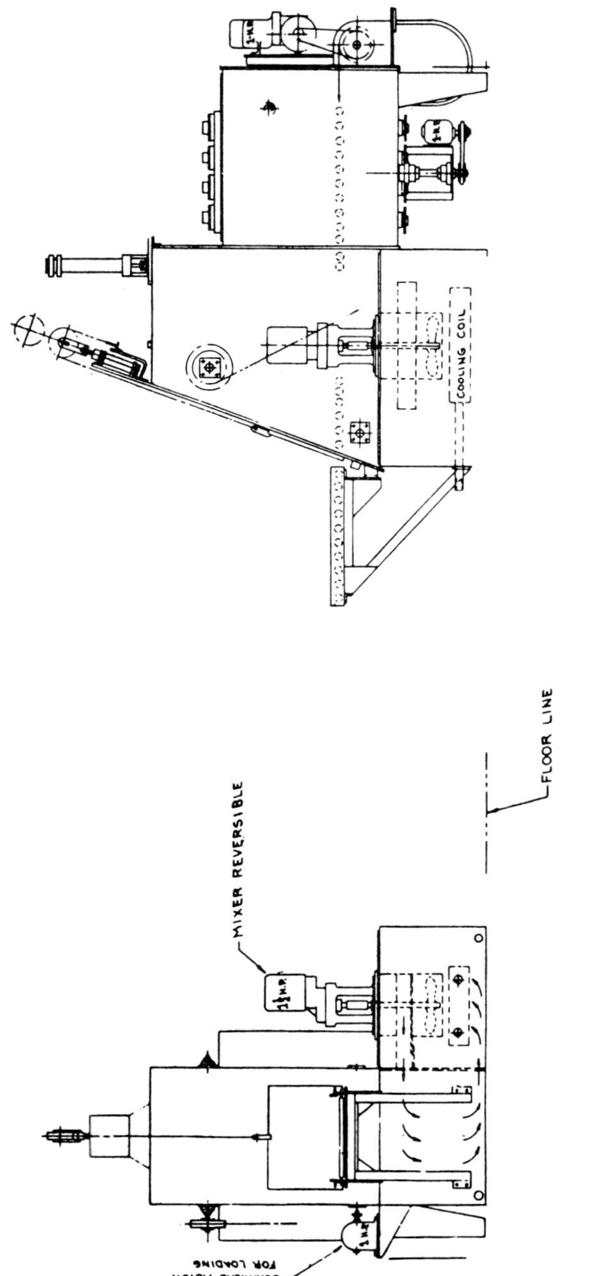

FIGURE 17B.

Schematic diagram of a vertical radiant-tube gas-fired automatic, batch type, hardening, carburizing, or carbonitriding furnace for heat treatment of sintered steel parts.

FIGURE 17C.
Installation of a vertical, radiant-tube, automatic batch, carbonitriding furnace and endothermic gas generator, for case hardening small, sintered iron bushings

17.2 STEAM-OXIDIZING OR BLUEING:

After iron parts have been sintered, it is sometimes desirable to impart a steam-oxidized surface or blueing effect. This is done to certain iron parts for wear resistance, color, or corrosion resistance. To obtain the best corrosion resistance, the parts should be tumbled in an oil-impregnated sawdust or quenched into soluble oil.

A steam tempering furnace is shown in figure 17D. The furnace usually is of the electric 100% forced-convection type constructed gas-tight to hold a steam atmosphere. In operation, the parts to be treated are loaded in the 500° F (260° C) furnace with the steam turned off. After the parts are preheated, the steam is turned on. The furnace is allowed to purge for about 30 minutes to remove the air. The temperature is raised to 1000° F (538° C) and the load allowed to soak out approximately 30 minutes or more. The load of parts can then be quenched into soluble oil for a deep blue-black color where oil impregnation is desired, or allowed to drop to 800° F (427° C) and then taken out to cool in air. The finish of the air-cooled parts is not as dark in color as the oil-treated parts.

FIGURE 17D.
A batch-type, 100% forced-convection steam tempering furnace for oxidizing, blueing of sintered iron parts for wear resistance, color, or corrosion resistance

17.3 VACUUM HEAT TREATING:

Neutral hardening and carburizing of P/M parts may also be accommodated in a partial pressure environment.

Carburizing in a partial pressure is accomplished by introducing a hydrocarbon gas (natural gas, CH_4) into a heated vacuum furnace.

This gas produces a carburizing atmosphere that varies in strength from zero to the saturation point. Work processed under these conditions exhibits an absorption of carbon characteristic to its behavior at a specific temperature. The potential of the atmosphere is a function of the partial pressure of the carbon and not the total pressure of the system. When the carburizing gas is admitted, carburization is instantaneous. Absorption of carbon immediately ceases upon the evacuation of the carburizing gas. The rate of carbon absorption increases with the increase of partial pressure of carbon. The maximum rate of carbon absorption occurs at the saturation point of carbon vapor at a specific temperature. For instance, low carbon steel heated to 1900° F (1038° C) and subjected to a saturated carbon will swiftly absorb carbon to a saturated surface carbon concentration.

A typical partial pressure neutral hardening and carburizing furnace is shown in figures 17E and 17F.

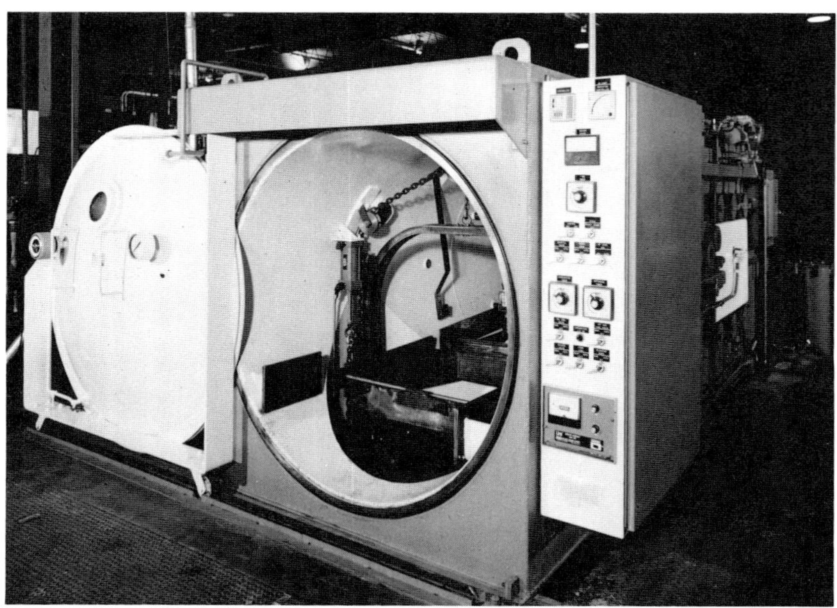

FIGURE 17E.

Installation of typical partial-pressure neutral hardening and carburizing furnace

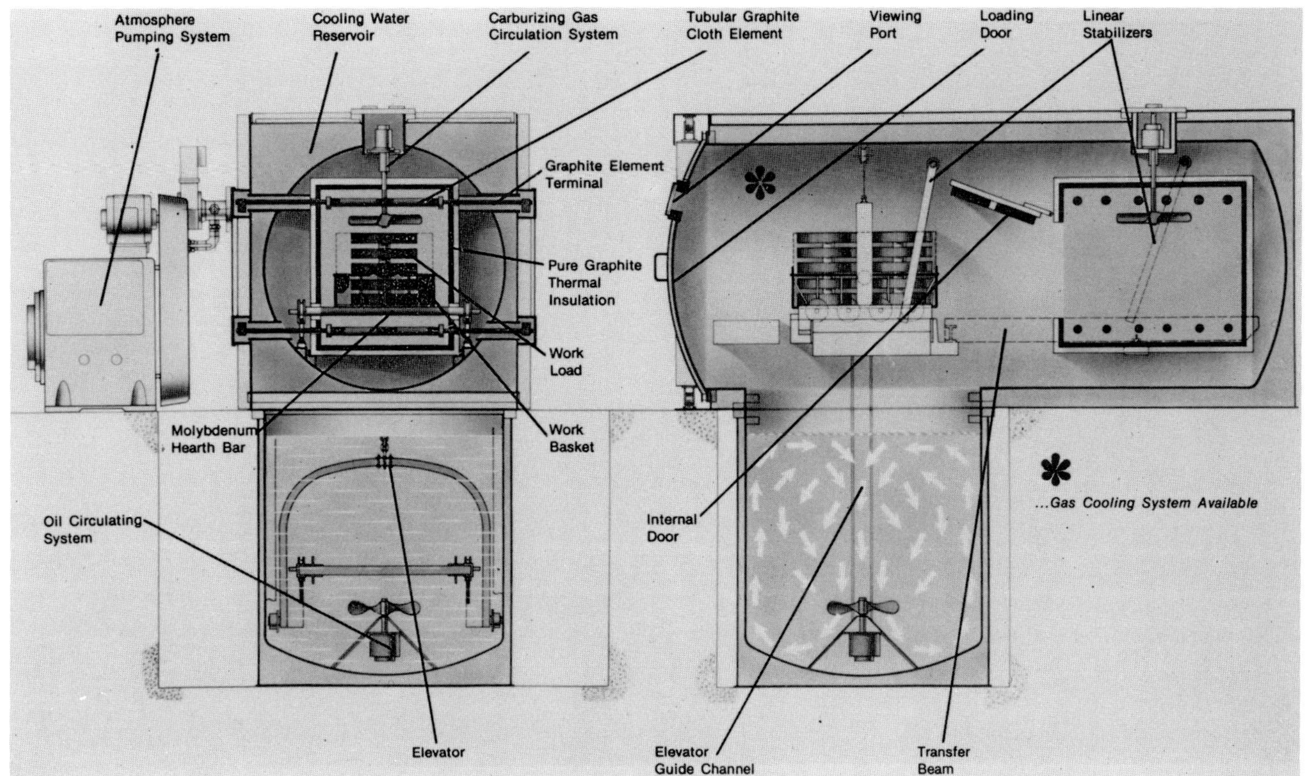

FIGURE 17F.

Cross section (left) and longitudinal section (right) of typical partial-pressure neutral hardening and carburizing furnace

18.0 TEMPERATURE CONTROL IN SINTERING FURNACES

Instrumentation plays an important part in obtaining the temperature control required for good sintering. Two general types of control are widely used: on-off control and proportioning control.

In connection with either of these methods, the proper sensing device (usually a thermocouple)must be used.

In addition to the control thermal element, some type of thermocouple protective device should always be installed. This will prevent run-away furnace temperatures in the event of thermocouple failure. However, do not over instrument; this can lead to high maintenance costs.

18.1 ON-OFF CONTROL:

As shown in figure 18A, the primary temperature-sensing element is a thermocouple extending through the furnace roof or wall into the heating chamber. A difference in temperature between the hot end and cold end of the thermocouple induces an electric millivolt output from the thermocouple, which varies with temperature. This tiny electric signal is conducted to a temperature-control instrument, often called a "pyrometer", which in turn

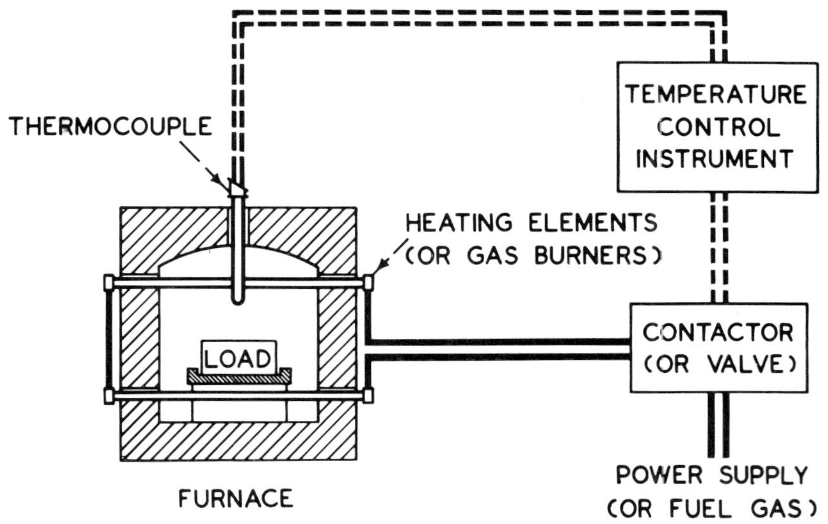

FIGURE 18A.

Schematic arrangement of elements used for a typical on-off temperature-control system

controls operation of a contactor, or valve (either motor-operated or solenoid-operated), commonly for on-and-off input of power or combustion air. Or, as discussed below, with other suitable accessories energy flow is continuous, but throttled in greater or lesser amounts proportional to demands. Long or high furnaces often have several thermocouples and "zones" of energy control. Zoning helps apply heat where most needed, such as in heating-up areas, and hold back heat input where little needed, such as in soaking zones.

18.2 PROPORTIONING CONTROL:

Proportioning controllers are of the position-adjusting type, duration-adjusting type or current-adjusting type. The position-adjusting type is used with gas-fired furnaces; the duration-adjusting type and current-adjusting type are used on electric furnaces. The controllers develop proportioning type of control by incorporating proportional, reset, and rate actions. These three actions closely regulate temperature by continuously throttling heat input according to the speed, size, and duration of temperature changes.

18.2.1 POSITION-ADJUSTING TYPE CONTROL:

a. Proportional action adjusts valve position to balance heat supply against distance temperature is off set-point. This eliminates cycling, but without the reset action described below will allow the controlled temperature to wander with load changes.

b. Reset action acts, whenever temperature is off set point, to correct heat supply in proportion to length of time it is away from desired value. This eliminates offset after a load change by continuously adjusting heat supply until temperature returns to its set-point.

c. Rate action acts whenever temperature is changing, to adjust heat supply in proportion to the speeds at which the change is occurring. This shortens the time needed to return the process to set-point, and eliminates or reduces overshoot after an upset.

18.2.2 DURATION-ADJUSTING TYPE CONTROL:

In the duration-adjusting type of proportional controllers, the proportional reset and rate actions adjust heat input by regulating the successive durations of on-time to off-time according to the size, speed and duration of temperature changes.

a. Proportional action balances the ratio of on-off time against the distance temperature is off set point.

b. Reset action acts as described in 18.2.1b

c. Rate action acts as described in 18.2.1c

18.2.3 CURRENT-ADJUSTING CONTROL, WITH MAGNETIC AMPLIFIER AND SATURABLE-CORE REACTOR:

With this type control, the three control functions are the same as described above under positioning-type control, excepting that the magnetic amplifier and reactor replace the valve.

a. A saturable-core reactor is essentially a set of coils wound on a transformer core having a variable impedence to regulate current flow. (This could be called an electric valve.) The reactor is connected in series with the load, and by means of direct current in the control winding varies the current in the load.

b. With the addition of a standard temperature measuring instrument and a current-adjusting type control feeding into a magnetic amplifier, the furnace can be controlled to a very narrow temperature band.

An advantage of a saturable-core-reactor type control is that the input to the furnace will match the demand. Also, the stepless-control feature of a reactor prolongs element life. Overheating is prevented because the system maintains the lowest possible temperature gradient between the furnace and material being heated.

When applying a saturable-core reactor to a molybdenum furnace, the current-limiting system should be used to prevent overloading the heating element on start-up of a cold furnace, where the cold resistance is only one-fifth to one-tenth the hot resistance. The automatic current limiter provides overload protection for the molybdenum. A schematic diagram of this system is shown in figure 18B. When full load current is exceeded, the current

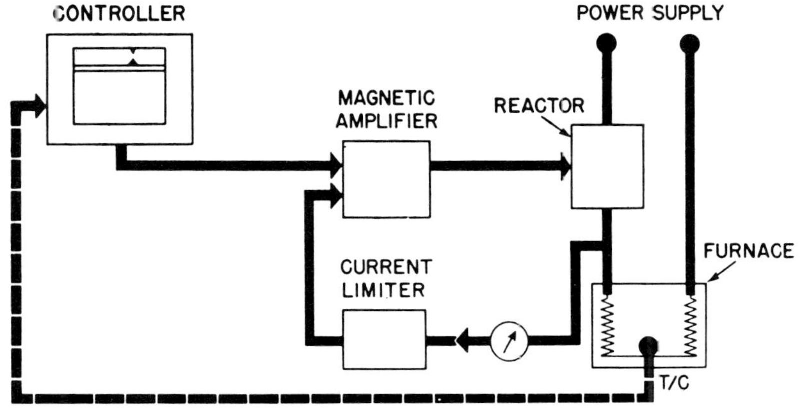

FIGURE 18B.

Principles of a current-limiting system used in conjunction with a saturable-core reactor to provide unattended, safe start-up for molybdenum heating elements

limiter feeds enough signal to the magnetic amplifier to maintain the load current approximately at its full load value. This brings the furnace up to temperature automatically and unattended in a safe and rapid manner.

A disadvantage of a saturable-core control system is the higher initial cost of the equipment compared to the cost of on-off control or duration-adjusting type control. The saturable-core reactor with current-limiting system is desirable on molybdenum furnaces as previously stated, to provide fast start-up or fast heat cycling of a furnace. Saturable-core reactor systems are used also on other types of heating elements because of the better temperature control, and the elimination of contactors. This system can also increase heating element life materially.

There are a variety of instrumentations available for furnaces, but the above are the basic principles on which they operate. Proportioning control is used in conjunction with a standard pyrometer whenever straight line temperature control is required.

18.3 THERMOCOUPLES:

The number of metals and alloys suitable for use as thermocouples on a commercial basis is limited. At high temperatures the materials must be resistant to oxidation, recrystallization, melting, and contamination by reducing atmospheres. They must develop an emf high enough to be measured without the use of delicate instruments, be of reasonable cost, be readily obtainable in uniform quality, and be reproducible.

There is a definite temperature limit for the different commercially-available thermocouples in different atmospheres. It is important, therefore, that the proper type couple be used for the desired temperature range (Table 18.1).

TABLE 18.1
TYPES OF THERMOCOUPLES

TYPE	TEMPERATURE RANGE
Copper - Constantan	-300° - +600° F (-218° - +317° C)
Iron - Constanan	0° - 1400° F (-18° - 760° C)
Chromel - Alumel	0° - 2000° F (-18° - 1093° C)
Nickel-Nickel Moly	0° - 2400° F (-18° - 1316° C)
Platinum-Platinum Rhodium	0° - 2800° F (-18° - 1538° C)
Tungsten-Tungsten Rhenium	1000° - 4300° F (+538° - 2371° C)

In sintering applications where the operating temperature range is up to 2000° F (1093° C) for nonferrous and ferrous applications, chromel-alumel thermocouples are used. Also, they may be used up to 2400° F (1315° C) for intermittent use. In applications where sintering is done above 2000° F (1093° C), such as for iron P/M parts, radiation pyrometers and platinum-platinum rhodium thermocouples are used.

Above 2500° F (1370° C), radiation-type detectors or tungsten-tungsten rhenium thermocouples generally are used. Bare nickel-nickel moly thermocouples have been used with some degree of success up to temperatures of 2400° F (1316° C) in vacuum furnaces. Nickel-nickel moly is the only known thermocouple material that will withstand a vacuum environment without a protection tube.

18.3.1 CHROMEL-ALUMEL:

The chromel-alumel couple uses chromel as the positive element and alumel as the negative element. Chromel is an alloy composed of 90% nickel and 10% chromium. Alumel is an alloy of 94% nickel, 2% aluminum, 3% manganese, and 1% silicon.

The recommended wire sizes, B and S gauge, for chromel-alumel couples in sintering furnaces, are 8- and 14-gauge. Thermocouple response is faster with the lighter gauge wire; however, the heavier wire resists deterioration.

Chromel-alumel couples have been found to be reasonably stable thermoelectrically when heated in a clean oxidizing atmosphere. When the couple is exposed to a reducing atmosphere containing wet hydrogen or carbon monoxide, it becomes contaminated, and the output is altered. It is, therefore, necessary to protect the thermocouple from the reducing gases. This is done by use of a protection tube, vented with circulating air, nitrogen, argon, or helium. Thermocouple protection tubes can develop leaks, and therefore, couples should be checked frequently. Furthermore, a second thermocouple connected to an excess-temperature cut-off should be employed as a safety measure.

Protection tubes for use with chromel-alumel thermocouples for temperatures up to 2000° F (1093° C) are made of heat-resisting alloys, such as nickel-chrome or nickel-chrome-iron. In clean atmospheres such as hydrogen or dissociated ammonia, thin-wall tubing of 80 nickel-20 chrome usually is used. Where the atmosphere is carburizing, such as low dew point endothermic gas with or without enrichment, protection tubes of heavy-wall Inconel pipe are used. Although the response to temperature change is slower, longer life of tube is the compensating factor. Protection tubes of lower alloys such as 35-15 and 18-8 are being used in applications where atmosphere is clean and temperatures are in the lower range.

For vacuum applications up to 2000° F (1093° C) chromel-alumel thermocouples generally are embedded in magnesium or aluminum oxide in Inconel sheaths.

18.3.2 NICKEL-NICKEL MOLY:

Bare nickel-nickel moly thermocouples have been used with some success in vacuum furnaces up to temperatures of 2400° F (1316° C). This same type of thermocouple has been used with the proper protective-sheath material sealing the thermocouple from the furnace atmosphere and also air for temperatures up to 2400° F (1316° C). The source of the thermocouple wire has been limited and for this reason nickel-nickel moly thermocouples have not been used universally, although the thermocouples are less costly than platinum-platinum rhodium.

18.3.3 PLATINUM-PLATINUM RHODIUM:

Platinum-platinum rhodium couples are called noble-metal thermocouples. They are made of pure platinum negative elements, and the positive element of either 87% platinum, 13% rhodium or 90% platinum, 10% rhodium.

"Platinum" thermocouples are easily contaminated. Hydrogen is especially detrimental to platinum, and will ruin a couple after even short exposure above 2000° F (1093° C), and even at lower temperatures, though at a slower rate. Silicon and other metallic vapors will also contaminate the "platinum" thermocouple, and therefore, protection tubes are always required. The couples usually are protected with a double tube. First, multiple ceramic insulators are used over the wires. This assembly is then inserted into a small-diameter ceramic thin-wall tube; this tube then is put into a larger-diameter ceramic tube.

The material used for protection tubes is a highly refractory porcelain which should be glazed to be impervious to gases.

"Platinum" thermocouples normally are limited to use in the temperature range from 1000° - 2500° F (538° - 1371° C). They can be used up to 2800° F (1538° C) with shorter life. They are commonly employed in sintering furnaces operating in the range from 1850° - 2100° F (1010° - 1150° C) for iron P/M parts. Above 2000° F (1093° C), the inner protection tube should be vented with air, nitrogen, argon, or helium to minimize hydrogen contamination. "Platinum" couples are usually not employed for temperatures under 1800° F (982° C) because the use of double ceramic protection tubes causes sluggish control in the lower-temperature range.

For vacuum applications up to 2800° F (1538° C), platinum-platinum rhodium thermocouples generally are embedded in magnesium or aluminum oxide in either molybdenum or tantalum sheaths.

18.3.4 TUNGSTEN-TUNGSTEN RHENIUM:

Tungsten-tungsten rhenium thermocouples now are available for use in hydrogen atmospheres or vacuum, with proper sheath material, up to temperatues of 4300° F (2371° C). Either tungsten or tantalum can be used for sheath material in vacuum. Tungsten must be used as the sheath material in a hydrogen atmosphere because tantalum becomes embrittled with hydrogen. The sheath material must be sealed to prevent air from entering from the terminals.

18.3.5 RADIATION-TYPE TEMPERATURE DETECTOR:

The radiation-type temperature detector utilizes the principle of the Stefan-Boltzmann law of radiant energy, which states that the intensity of radiant heat emitted from the surface of a body increases as the fourth power of its absolute temperature.

Radiation-type detectors are used with high-temperature sintering furnaces, such as the molybdenum-element type with hydrogen or dissociated ammonia atmospheres, because they measure without coming into direct contact with the atmosphere or the high temperature. These units sight on closed-end target tubes or into the furnace to read temperature. They are used sometimes above 2000° F (1093° C) and commonly above 2500° F (1371° C).

19.0 FURNACE ATMOSPHERES AND ATMOSPHERE GAS PRODUCERS

Furnace protective atmospheres reduce oxides, prevent the formation of oxides, and/or prevent decarburization of powders during manufacture, or in P/M compacts during sintering. Protective atmospheres play a very important part in the field of powder metallurgy. Following is a discussion of the most common types of gases utilized for protective atmospheres, properties of the gases, how to purify the gases, advantages and disadvantages, applications, how the gases are made, and how the gas compositions or impurities are determined. Gas compositions and costs are summarized in Table 19.1 and applications in Table 19.2

19.1 HYDROGEN:

At elevated temperatures hydrogen (H_2) is highly reducing to oxides of some metals, such as iron and copper. This property, combined with the ready availability of H_2 has been largely responsible for its wide use.

Hydrogen is very flammable, having an extremely high rate of flame propagation Although this explosive property requires that it be handled with respect, it has been used in large production furnaces for many years with a good safety record. Because of its high rate of flame propagation, hydrogen burns with a short, hot flame immediately upon contact with air. The flame is an almost colorless blue.

This gas is very light in weight—in fact, it is the lightest element known. Its specific gravity is only 0.069 as compared to 1.0 for air. It is easily displaced by air, and rushes out the top of the furnace door openings rapidly when free to do so.

Hydrogen is an excellent heat conductor. The thermal conductivity is seven times that of air. Hence it accelerates both the heating and cooling rates of the work in furnaces. The thermal losses in furnaces are higher with hydrogen atmospheres than when using heavier, less conductive gases.

Hydrogen is made in a variety of ways, and generally contains fractional percentages of impurities in the order of 0.1 to 0.5%, which may or may not be troublesome. Impurities of oxygen and water vapor can be removed readily if troublesome. Impurities of carbon monoxide and methane result from some methods of manufacture, but attempts are seldom made to remove them.

The most accepted method at present to remove oxygen is to pass the gas through a palladium-catalyst purifier, which forms water. The unit operates at room temperature and is very effective. It lowers the oxygen to less than one part per million parts. Hydrogen delivered from this purifier contains water vapor, of course; and if dry gas is necessary, moisture removal is required.

Hydrogen as manufactured generally is saturated with water vapor. The generators are very expensive, and are found only in plants having high consumption. More commonly, hydrogen is purchased in cylinders or trailer loads of large steel tubes at pressures of 1800 to 2400 psi. If the gas is purchased to a dryness specification, it is pre-dried before bottling and will always stay dry. It is not uncommon, however, to find it quite dry when first drawing it from a high-pressure cylinder, but as it is used and the pressure drops, the moisture content increases.

Moisture removal is customarily achieved by refrigeration or by adsorption—depending upon the moisture content and volume of gas used. Line hydrogen from an in-plant producer, as contrasted to bottled gas, commonly is saturated with water at room temperature and at atmospheric presure. At high rates of flow, say one thousand (M) cfh (8.81/s or more, the gas first can be refrigerated to +40°F (4°C). This will remove about two-thirds of the water vapor.

For eliminating the remaining moisture in the two step procedure, or for drying all the way in one step, regenerative adsorption-type driers are used. Such driers usually contain beds of desiccant materials in one or two towers, depending on whether service is to be intermittent or continuous. A common desiccant is activated alumina, which will dry the gas very effectively. For critical applications requiring the driest possible gas, a material called "molecular sieve" is highly successful.

After a tower has adsorbed its full capacity of water in the capillary pores of the desiccant, the bed is heated and the air or gas is circulated through it to reactivate the desiccant. The bed is then cooled, and is ready for use again.

Typical applications of hydrogen, sometimes unpurified, sometimes purified, in the manufacture of powders, are in the: redution of oxides of iron, molybdenum, tungsten, cobalt, nickel, and 18-8 chromium-nickel stainless stee; annealing of electrolytic- and carbonyl-iron powders, and carburizing of tungsten powders in lamp black to form tungsten carbide. In the sintering field hydrogen is used for sintering compacts of molybdenum, tungsten, tungsten carbides, stainless steels, brass, aluminum and aluminum and aluminum alloys, etc.

TABLE 19.1
APPROXIMATE COMPOSITIONS AND COSTS OF TYPICAL PROTECTIVE ATMOSPHERE GASES

Atmosphere Gases	Approximate Air: Natural Gas Ratio	Dew Point °F Entering Furnace	Properties At Elevated Temperatures	Composition, Percent								Cost Per M (b) Cubic Feet (Dollars)
				Carbon Dioxide CO_2	Oxygen O_2 (i)	Carbon Monoxide CO	Hydrogen H_2	Methane CH_4	Nitrogen N_2	Argon A	Helium He	
1 Hydrogen												
A. By Electrolysis of Water												
1. Direct from Cells												
a. Unpurified—Saturated		+70 to +90	R, r, D3	0.0	0.2	0.0	99.8	0.0	0.0	0.0	0.0	3.50
b. Purified (e)		-80 or lower (k)	R, r, N	0.0	0.0	0.0	100.0	0.0	0.0	0.0	0.0	3.64
B. From Bottles												
1. Unpurified		-30 to +20	R, r, D1	0.0	0.0	0.0	99.95	0.0	0.0	0.0	0.0	32.00 to 45.00 (h)
2. Purified (e)		-80 or lower (k)	R, r, N	0.0	0.0	0.0	100.0	0.0	0.0	0.0	0.0	32.08 to 45.08 (h)
C. By Catalytic Conversion of Hydrocarbons (c)												
1. Unpurified—Saturated		+70 to +90	R, r, D3	0.05	0.0	0.02	99.73	0.20	0.0 (d)	0.0 (d)	0.0 (d)	3.70
2. Dried		-100 or lower	R, r, N	0.002	0.0	0.02	99.97	0.01	0.0 (d)	0.0 (d)	0.0 (d)	4.20
D. From Liquid Hydrogen		-84 or lower	R, r, N	0.0	0.0	0.0	99.998	0.0	0.0	0.0	0.0	10.00 to 12.00 (h,l)
2 Nitrogen—Bottled		-64 or lower	O, o, D1	0.0	0.0	0.0	0.0	0.0	99.996	0.0	0.0	20.00 to 32.00 (h)
3 Hydrogen-Nitrogen Mixtures												
A. Dissociated Ammonia												
1. As Reacted—Dry		-60 to -40	R, r, N	0.0	0.0	0.0	75.0	0.0	25.0	0.0	0.0	4.14 to 7.91 (h)
2. Moisture Added—Saturated		+70 to +100	R, r, D3	0.0	0.0	0.0	75.0	0.0	25.0	0.0	0.0	4.34 to 8.11 (h)
B. Burned Dissociated Ammonia												
1. Rich—Saturated (f)		+70 to +90	R, r, D3	0.0	0.0	0.0	24.0	0.0	76.0	0.0	0.0	3.02 to 5.78 (h)
2. Lean—Saturated (g)		+70 to +90	O, n, D3	0.0	0.0	0.0	1.0	0.0	99.0	0.0	0.0	2.51 to 4.80 (h)
C. Direct Catalytic Conversion of Ammonia and Ai												
1. Rich												
a. As Reacted—Saturated		+70 to +90	R, r, D3	0.0	0.0	0.0	25.0	0.0	75.0	0.0	0.0	2.97 to 5.78 (h)
b. Cooled to 40°F		+40	R, r, D2	0.0	0.0	0.0	25.0	0.0	75.0	0.0	0.0	2.97 to 5.78 (h)
c. Dried (e)		-80 or lower (k)	R, r, N	0.0	0.0	0.0	25.0	0.0	75.0	0.0	0.0	3.05 to 5.86 (h)
2. Lean												
a. As Reacted—Saturated		+70 to +90	O, n, D3	0.0	0.0	0.0	1.0	0.0	99.0	0.0	0.0	2.43 to 4.72 (h)
b. Cooled to 40°F		+40	O, n, D2	0.0	0.0	0.0	1.0	0.0	99.0	0.0	0.0	2.43 to 4.72 (h)
c. Dried (e)		-80 or lower (k)	N, n, N	0.0	0.0	0.0	1.0	0.0	99.0	0.0	0.0	2.51 to 4.82 (h)
4 Reformed Hydrocarbon Gases												
A. Exothermic Gas												
1. Rich												
a. As Reacted—Saturated	6:1	+70 to +90	R, r, D3	5.0	0.0	10.0	14.0	1.0	70.0	0.0	0.0	0.33
b. Refrigerated	6:1	+40 to +60	R, r, D2	5.0	0.0	10.0	14.0	1.0	70.0	0.0	0.0	0.39
2. Medium Rich—Saturated	6.75:1	+70 to +90	O, r, D3	6.3	0.0	8.8	11.5	0.3	73.1	0.0	0.0	0.33
3. Lean—Saturated	10.25:1	+70 to +90	O, n, D3	11.5	0.0	0.7	0.7	0.0	87.1	0.0	0.0	0.41
B. Purified Exothermic Gas												
1. Rich	6:1	-40 or Lower	R, r, C1	0.0	0.0	10.8	15.0	1.0	73.2	0.0	0.0	0.73
2. Medium Rich												
a. As Reacted	6.75:1	-40 or Lower	R, r, C1	0.0	0.0	9.0	12.0	0.2	78.8	0.0	0.0	0.73
b. Methane Added (j)	6.75:1	-40 or Lower	R, r, C3	0.0	0.0	8.5	11.4	5.2	74.9	0.0	0.0	0.78
3. Lean	10.25:1	-40 or Lower	N, n, N	0.0	0.0	0.7	0.7	0.0	98.6	0.0	0.0	0.83
C. Endothermic Gas												
1. Rich—Dry	2.4:1	-10 to +10	R, r, C3	0.0	0.0	20.0	38.0	0.5	41.5	0.0	0.0	0.52
2. Rich—Fairly Dry												
a. As Reacted	2.6:1	+20 to +30	R, r, C2	0.0	0.0	19.0	37.0	0.3	43.7	0.0	0.0	0.52
b. Methane Added (j)	2.6:1	+20 to +30	R, r, C3	0.0	0.0	18.1	35.2	5.3	41.4	0.0	0.0	0.57
3. Medium Rich—Saturated	3.5:1	+70 to +90	R, r, D1	1.2	0.0	16.5	27.6	0.0	54.7	0.0	0.0	0.55
4. Lean—Saturated	4.5:1	+70 to +90	R, r, D3	3.0	0.0	13.8	21.5	0.0	61.7	0.0	0.0	0.57
5 Argon—Bottled		-73 or Lower	N, n, N	0.0	0.0	0.0	0.0	0.0	0.0	99.996	0.0	65.00 to 95.00 (h)
6 Helium—Bottled		-73 or lower	N, n, N	0.0	0.0	0.0	0.0	0.0	0.0	0.0	99.995	100.00 to 115.00 (h)
7 Vacuum (Below 150 Microns)		—	R, r, N	0.0	0.0	0.0	0.0	0.0	0.0	0.0	0.0	—
8 Air												
A. Normal		+50 or Lower	O, o, D3	0.0	21.0	0.0	0.0	0.0	78.1	0.9	0.0	—
B. Wet		Above +50	O, o, D3	0.0	21.0	0.0	0.0	0.0	78.1	0.9	0.0	—

Foot Notes to Table 19.1

(a) Properties at elevated temperatures with:

Iron or iron oxides:	0-Oxidizing N-Neutral R-Reducing
Copper or copper oxides:	o-Oxidizing n-Neutral r-Reducing
Iron or iron-carbon:	C1-Mildly carburizing C2-Average carburizing C3-Strongly carburizing $\mathcal{N}$-Neutral D1-Mildly carburizing D2-Average decarburizing D3-Strongly decarburizing

(b) Costs include only materials consumed, such as power, gas and water. They do not include factors for obsolescence, installation, maintenance, labor, etc. The estimates have been based on the following unit costs as of November 1, 1976:

Power	2 cents per KWHR
Natural Gas	$1.00 per M cu. ft.
Water	$1.00 per M gal.
Ammonia	18 or 35 cents per lb. (tank truck or cylinder lots, respectively) (h)
Steam	$3.00 per M lb.

(c) Composition varies with purification stages utilized.

(d) Some traces will be present if contained in fuel gas processed.

(e) Purified at point of use.

(f) Can be refrigerated to 40° F to compare with 3Clb, and dried to –80 F or lower to compare with 3C1c.

(g) Can be refrigerated to 40° F to compare with 3C2b, and dried to –80 F or lower to compare with 3C2c.

(h) Transportation costs are included.

(i) Attention is directed to the fact that in atmosphere gases such as hydrogen, nitrogen, argon and helium, there are usually at least 10 ppm oxygen, and in some cases more, unless special precautions are taken.

(j) For carbonitriding, up to both 5% methane and 10% ammonia are added to the carrier gas.

(k) Dew points of –80 F or lower are achieved with molecular sieve driers having "closed-circuit reactivation," resulting in the approximate atmosphere-gas costs tabulated. Minus 100 F or better can be attained, however, by using driers with "bleed reactivation," in which dried gas is bled to waste through the bed being reactivated. This results in lower equipment costs but higher atmosphere-gas costs.

(l) This liquid-hydrogen price is based on using 750,000 to 300,000 cu. ft. hydrogen per month.

TABLE 19.2

POWDER METALLURGY APPLICATIONS AND RECOMMENDED ATMOSPHERE GASES(a)

Atmosphere Gases (b)	Annealing					Carburizing		Heat Treating		Infiltration		Reducing Oxides							Sintering																								
Material	Copper	Iron—Carbonyl	Iron—Electrolytic	Steels—Low Carbon	Steels—Medium Carbon	Iron	Tungsten (with lamp black)	Steels-Carbon	Steels-Copper	Iron	Steels-Carbon	Cobalt	Iron	Molybdenum	Nickel	Steels—Carbon & Alloy	Steels—Stainless	Tungsten	Alinico	Beryllium	Brass	Bronze	Carbides of Refractory Metals	Copper	Iron	Iron-Copper	Aluminum	Metal-Ceramic Combinations (Cermets)	Molybdenum	Nickel	Niobium (Columbium)	Silver	Steels—Carbon & Alloy	Steels—Stainless	Tantalum	Titanium	Thorium	Tungsten	Tungsten Alloys	Uranium	Uranium Oxide	Vanadium	Zirconium
1 Hydrogen																																											
A. By Electrolysis of Water																																											
1. Direct from Cells																																											
a. Unpurified—Saturated	●	●	●	●						●			●		●			●				●		●	●	●				●		●									●		
b. Purified							●					●		●			●		●		●		●				●		●					●				●	●	●			
B. From Bottles																																											
1. Unpurified	●	●	●	●						●			●		●			●				●		●	●	●				●		●									●		
2. Purified							●					●		●			●		●		●		●				●		●					●				●	●	●			
C. By Catalytic Conversion of Hydrocarbons																																											
1. Unpurified—Saturated	●	●	●	●						●			●		●			●				●		●	●	●				●		●									●		
2. Dried							●					●		●			●		●		●		●						●					●				●	●	●			
D. From Liquid Hydrogen							●					●		●			●		●		●		●				●		●					●				●	●	●			
2 Nitrogen—Bottled	●																										●																
3 Hydrogen-Nitrogen Mixtures																																											
A. Dissociated Ammonia																																											
1. As Reacted—Dry	●	●	●	●			●			●		●	●	●	●		●	●			●	●		●	●	●	●		●	●		●		●				●	●		●		
2. Moisture Added—Saturated																																											
B. Burned Dissociated Ammonia																																											
1. Rich—Saturated	●	●	●	●						●												●		●	●	●				●		●									●		
2. Lean—Saturated	●																																										
C. Direct Catalytic Conversion of Ammonia and Air																																											
1. Rich																																											
a. As Reacted—Saturated	●	●	●	●						●												●		●	●	●				●		●									●		
b. Cooled to 40°F																																											
c. Dried																	●				●						●							●									
2. Lean																																											
a. As Reacted—Saturated	●																																										
b. Cooled to 40°F																																											
c. Dried																																											
4 Reformed Hydrocarbon Gases																																											
A. Exothermic Gas																																											
1. Rich																																											
a. As Reacted—Saturated	●			●						●												●		●	●	●				●		●											
b. Refrigerated																																											
2. Medium Rich—Saturated	●			●																		●								●		●											
3. Lean—Saturated																																											
B. Purified Exothermic Gas																																											
1. Rich	●			●	●			●	●		●					●					●	●		●	●	●				●		●	●										
2. Medium Rich																																											
a. As Reacted	●			●	●			●	●	●	●					●					●	●		●	●	●				●		●	●										
b. Methane Added						●		●	●																																		
3. Lean	●																								●							●											
C. Endothermic Gas																																											
1. Rich—Dry	●				●	●		●	●		●					●					●	●		●									●										
2. Rich—Fairly Dry																																											
a. As Reacted	●				●			●	●	●	●					●					●	●		●									●										
b. Methane Added						●																																					
3. Medium Rich—Saturated	●			●						●												●		●	●	●				●		●											
4. Lean—Saturated	●			●						●												●		●	●	●				●		●											
5 Argon—Bottled	●																			●			●				●	●	●		●			●	●	●	●	●	●	●	●	●	●
6 Helium—Bottled	●																			●			●				●	●	●		●			●	●	●	●	●	●	●	●	●	●
7 Vacuum (Below 150 Microns)																				●			●		●	●	●	●	●		●		●	●	●	●	●	●	●	●	●	●	●
8 Air																																											
A. Normal																											●																
B. Wet																																											

(a) Where wet gases are recommended, dry gases will serve equally well, possibly better. Exceptions are in Section 4, wherein drier exothermic gases have higher carbon potential than wet gases, which may result in carburizing of low-carbon materials.
Applications of dry gases often require that the furnaces be built and operated in a manner that will contribute to holding high-purity atmospheres within the furnaces.

(b) For approximate compositions and costs of atmosphere gases see Table 19.1.

Six typical ways by which hydrogen is made for industrial purposes are:

1. Electrolysis of Distilled Water: Electric current passes through electrolyte in cells, causing hydrogen to collect at one electrode in each cell, and oxygen at the other. The electrolyte is either sodium hydroxide or potassium hydroxide in distilled water.

2. Electrolysis of Sodium Chloride Solution: Electric current passes through sodium chloride solution in cells, causing hydrogen to collect at one electrode and chlorine at the other, and forms sodium hydroxide in the electrolyte.

3. Catalytic Process for Converting Hydrocarbons: A hydrocarbon gas such as natural, propane, or butane gas is mixed with steam, heated, and passed through a catalyst to produce hydrogen plus carbon monoxide and a small amount of carbon dioxide. This mixture with steam is further catalyzed to convert carbon monoxide to carbon dioxide, accompanied by production of an additional volume of hydrogen equal to that of the carbon monoxide. The resulting mixture of hydrogen plus carbon dioxide is passed through a chemical absorbing tower which removes the carbon dioxide, leaving hydrogen and a slight amount of carbon monoxide as an impurity.

4. Catalytic Process for Converting Water Gas: Water gas (hydrogen plus carbon monoxide plus small amounts of carbon dioxide, nitrogen, and methane) is made by passing steam through hot coke. The carbon monoxide and carbon dioxide are then removed as described above, leaving the hydrogen.

5. Iron-Contact Process: Water gas made by passing steam through hot coke is partially burned with air in a tower containing trays of iron oxide; this heats the oxide and reduces it to iron. When the iron if fully reduced, the flow is shut off and steam is passed through the reduced iron, which produces hydrogen and iron oxides.

6. Liquid Hydrogen: Liquid hydrogen is cryogenically produced, using most any available feed hydrogen with its impurities, by compressing and cooling the gas in stages, and venting the impurities. The resultant product is 99.998% pure hydrogen which is vaporized as a gas. On-site storage vessels are used, which are periodically refilled by suppliers as the needs arise.

19.2 NITROGEN:

Nitrogen is non-flammable and is sometimes used as a purging gas for furnaces prior to and after use with flammable-gas atmospheres. At elevated temperatures, nitrogen is inert to iron and copper, but not inert to many of the reactive metals with which it forms nitrides. As a result nitrogen has found very little application in P/M work, except where it occurs naturally in the inexpensive reformed mixtures.

Nitrogen, made from air by liquefaction and rectification, is available in cylinders. Also it is produced with relatively small amounts of impurities from ammonia or from hydrocarbon gases.

19.3 HYDROGEN-NITROGEN MIXTURES:

Mixtures of hydrogen and nitrogen can be obtained from ammonia at lower cost than cylinder hydrogen. In many instances these mixtures serve equally as well as hydrogen, but sometimes they have unfavorable metallurgical reactions in nitriding the products. Therefore, each application is generally considered on its own merits.

19.3.1 DISSOCIATED AMMONIA:

Dissociated ammonia consists of 75% hydrogen-25% nitrogen. Normally, the dissociation or "cracking" of ammonia with heat, over a catalyst, is 99.95% or better complete. Thus there is only a trace (0.05% or less) of ammonia in the gas; oxygen is 0.0% and the moisture content is indicated by dew points of -40° to -60° F (-40° to -51° C).

Should the small amount of raw ammonia in the gas be objectionable, all traces can be eliminated by passing the gas through either: water, which will absorb ammonia but increase the moisture content (which can be removed); or activated alumina, or molecular sieve, which will adsorb ammonia, and also water, thus further drying the gas.

Because of its high hydrogen content, dissociated ammonia is highly reducing at elevated temperatures on various oxides. Also, it is highly flammable. Its specific gravity is 0.295 and thermal conductivity 5.507 compared to air, which is 1.0. Ammonia is generally available in cylinders containing 150 pounds (330 kg), in tank-truck deliveries of 2000 to 4000 pounds (4400 to 8800 kg), or in bulk tank-truck or railway tank-car deliveries into the purchaser's storage facilities, in lots of 26,000 and 52,000 pounds (57,200 to 114,400 kg).

The uses of dissociated ammonia overlap, to a great extent, those of hydrogen and some other low-cost gaseous mixtures. Typical applications of dissociated ammonia include: annealing electrolytic-iron powder, reducing stainless-steel 18 Cr-8 Ni powder, and sintering of brass, copper, iron, iron-copper, tungsten and tungsten alloys, aluminum and aluminum alloys, and stainless steels. Dissociated ammonia is sometimes avoided where molecular nitrogen causes nitriding and consequent hardness and embrittlement, such as in sintering stainless steel or molybdenum compacts.

A simplified flow diagram of an ammonia dissociator is shown in figure 19A. Liquid ammonia from the tank enters a vaporizer at high pressure, where heat converts the liquid into vapor. The pressure of the vapor is then reduced in an expansion valve and the low-pressure vapor passes through a dissociating element filled with catalyst. Here, within the heated reaction chamber held at 1650° - 1850° F (899° - 1010° C) the ammonia is dissociated into its constituents hydrogen and nitrogen. The gaseous mixture then passes from the outlet of the dissociating element to the furnace.

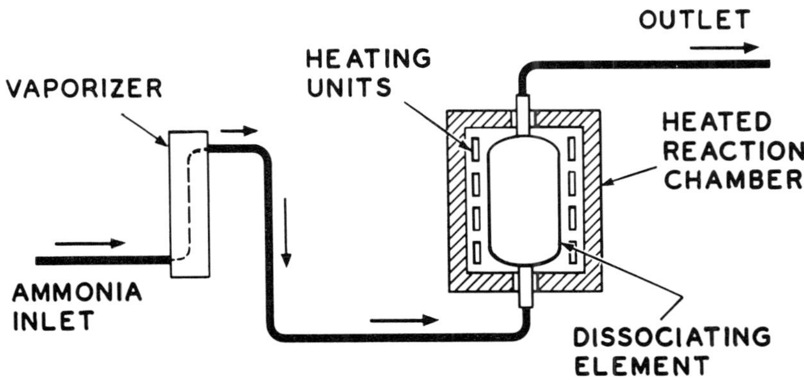

FIGURE 19A.
Simplified flow diagram of typical ammonia dissociator

19.3.2 BURNED DISSOCIATED AMMONIA:

Burning of dissociated ammonia pre-mixed with air, in a reaction chamber, gives low hydrogen in nitrogen, varying from 24% at the upper limit to 0.5% at the lower limit, depending upon the amount of air. The mixture is saturated with water, but it can be dried if necessary.

These low-hydrogen nitrogen mixtures are less highly reducing and less flammable than those with higher hydrogen. In fact, they are non-flammable when the hydrogen is at or near the lower limit.

The equipment in which dissociated ammonia is burned is basically the same as the exothermic gas producer shown in figure 19C, and described in the related text. In this instance, dissociated ammonia takes the place of the hydrocarbon fuel gas entering the system.

19.3.3 DIRECT CATALYTIC CONVERSION OF AMMONIA AND AIR:

Similar mixtures varying from 25% to 0.5% hydrogen, remainder nitrogen, are made by direct catalytic oxidation of ammonia with air, as compared to the two step process of dissociation plus combustion as just described.

Direct catalytic generators are available with various modifications. In a typical unit, ammonia vapor and air are mixed and fed into a catalytic chamber provided with a starting heater to hold 250° F (121° C) at the outset. Once the reaction is begun it gives off heat and is self-supporting. The hydrogen-nitrogen-ammonia mixture from this first-stage converter is combined with additional air and passed into a second-stage converter in which the reaction goes to completion. The mixed hydrogen-nitrogen gas then leaves the outlet and goes to the furnace. Some units use ammonia vapor, in which case the output gas is saturated with water. Some use liquid ammonia, the expansion of which, in a vaporizer, refrigerates the output gas to a dew point of about +40° F (4.4° C) and requires a smaller drier if dry gas is needed.

19.4 REFORMED HYDROCARBON GASES:

There is a family of low-cost gaseous mixtures made by reforming hydrocarbon gases which will be described in this section. Each type of gas is named from the kind of reaction by which it is produced, has wide usage in various types of powder metallurgy, and has its own advantages and limitations. The gases are called "exothermic gas", "purified exothermic gas", and "endothermic gas".

19.4.1 EXOTHERMIC GAS:

If air and fuel gas are mixed in proper proportions for complete combustion, and the mixture is burned in a refractory-lined reaction chamber, a water-jacket may be needed because so much heat is given off by the reaction. When heat is generated, it is termed an "exothermic reaction". Accordingly, for convenience in protective-atmosphere work, the products of combustion obtained in this manner are labeled as being "exothermic gas".

Figure 19B indicates, at the right, the composition, with the exception of nitrogen, of exothermic gas from the producer under various operating conditions. (Endothermic gas, at the left, is discussed later). After adding the percentage of the various constituents shown on the chart for any single operating condition (air: gas ratio), the difference between 100 and this sum is the percentage of nitrogen. In all cases, nitrogen is actually the largest single constituent. Percentages of the constituents can be read in the vertical column at the left of the chart. The numbers in the horizontal column at the bottom indicate the ratios of the input mixtures of air and natural gas, which is the most common fuel gas encountered. Each number indicates the volumes of air compared to one volume of 1000 Btu (252 Kg Cal) natural gas in the mixture, such as 6:1 at the center of the chart or 10:1 at the right. Other charts are available to show similar compositions when using different hydrocarbon fuel gases, such as coke-oven, propane, or butane gas.

As indicated at the extreme right of figure 19B, nearly complete combustion, or the point at which practically all combustibles in the fuel gas are burned up, takes place with a ratio of about 10.25 volumes of air to 1 volume of natural gas. The composition of the resulting output gaseous mixture from

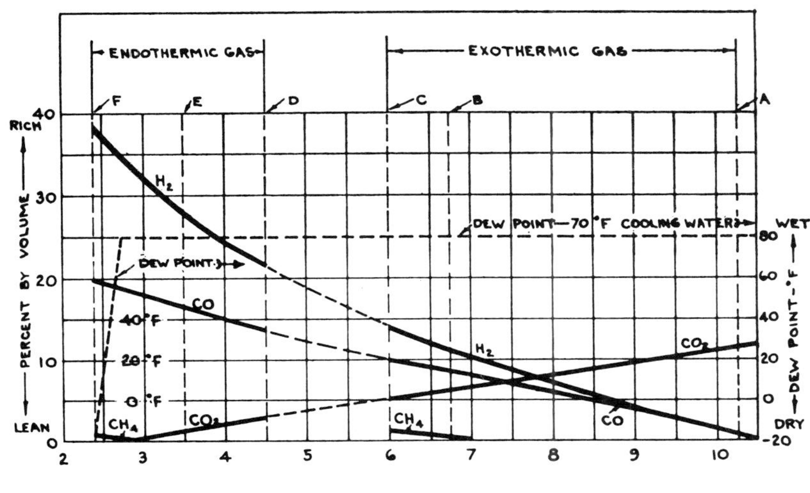

FIGURE 19B.

Constituents of endothermic gas (left) and exothermic gas (right), except nitrogen which is the remainder, at various ratios of air to natural gas

the exothermic gas produced, as indicated by the points at which the vertical dash-line A intersects the curves, is about 0.7% hydrogen, 0.7% carbon monoxide, 11.5% carbon dioxide, and 87.1% nitrogen (not shown on the chart). This gas is saturated with water vapor at about 10° F (6° C) above the temperature of the cooling water used in the surface cooler of the producer. In figure 19B, the dew point is indicated to be 80° F (27° C) with cooling-water temperature of 70° F (21° C), which is low compared with the high amount (about 18%) of moisture initially formed in the reaction. In other words, the cooler condenses out most of the mositure.

The complete combustion of natural gas, which is mostly methane, with air which is about 21% oxygen, 79% nitrogen, or 1 part oxygen to 3.8 parts nitrogen, is represented by the equation: CH_4 + air (2 O_2 + 7.6 N_2) = CO_2 + 2 H_2O + 7.6 N_2.

The gaseous mixture at vertical line A is relatively inert to some materials at elevated temperatures. For example, it will not oxidize hot copper, tin, or silver. It will, however, oxidize hot iron and the reactive metals because of the high percentages of carbon dioxide and water as opposed to the extremely low percentages of reducing components, hydrogen and carbon monoxide.

Being lean in fuel gas, this reformed mixture at A is called "lean exothermic gas". It is the lowest cost atmosphere gas available.

Actually, lean exothermic gas has very little application in powder metallurgy. Where it can be used, most individuals prefer to have a mixture richer in hydrogen to help overcome oxidizing effects of impurities which infiltrate into the furnace atmosphere.

When sintering bronze parts, for example, some users prefer to operate at about 6.5 or 7 to 1 air-gas ratio. As indicated in figure 19B, using a mean ratio of 6.75:1, at vertical dash-line B, this "medium-rich" mixture has a composition of about 11.5% hydrogen, 8.8% carbon monoxide, 6.3% carbon dioxide, 0.3% methane, and 73.1% nitrogen. This gas is sometimes refrigerated, with the belief that reduction of the water content to about +40° F (4.4° C) dew point will give better results on the product being treated.

The richest exothermic gas is obtained by still further reducing the air to a 6:1 air-gas ratio, as indicated by the vertical dash-line C at the center in figure 19B. Here the composition is about 14% hydrogen, 10% carbon monoxide, 1% methane, 5% carbon dioxide, and 70% nitrogen.

Carbon dioxide and water combine with and remove carbon from the surfaces of carbon-steel compacts being sintered. For example, CO_2 + C (free graphite) = 2 CO; and CO_2 + Fe_3C (carbon combined with iron) = 3 Fe + 2 CO. This removal of carbon is called surface decarburization. It results in the parts having reduced strength and soft skins subject to rapid wear. When sintered and hardened without decarburization, parts are hard to the surface and have good wear-resistance and strength.

When an atmosphere gas neither carburizes nor decarburizes a ferrous metal, the gas is said to have a "carbon potential" equal to the carbon content of the metal. For example, an atmosphere neutral to compacts with 0.40% carbon at the temperature involved, would have a carbon potential of 0.40% carbon.

Sometimes exothermic gas is dried to help improve its properties, but drying alone is generally considered to be not too helpful, because water can be formed in the furnace by the "water-gas reaction" between hydrogen and carbon dioxide; the equation for this is...$H_2 + CO_2 = CO + H_20$. Thus, to purify the gas effectively, carbon dioxide also should be removed, in order to obtain a stable, dry furnace atmosphere.

A hydrocarbon fuel gas occasionally contains sulphur, which commonly is the case with coke-oven gas. Generally the sulphur is removed from the exothermic gas before it goes into the furnace, by means of inexpensive purifiers, if the gas is to be used on nonferrous parts containing copper or silver. Sulphur removal is not necessary for iron parts.

Benefits are low flammability of the gas, which contributes to safety in its use; low thermal conductivity, which helps the operating economy of the furnace with respect to thermal losses; and relatively trouble-free operation of the exothermic gas producer.

Rich exothermic gas is useful for sintering copper, bronze, silver, iron-copper and iron. Its carbon dioxide and water contents, however, make it unsuitable for some P/M parts. Brass may lose zinc and become discolored and have inferior mechanical properties, and carbon steel may be decarburized. Steel parts may be recarburized, however, before or during heat treatment.

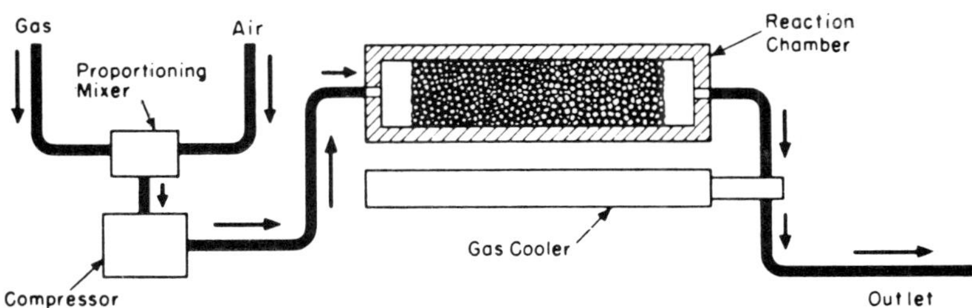

FIGURE 19C.
Simplified flow diagram of typical exothermic gas producer

Figure 19C shows a simplified flow diagram of a typical exothermic gas producer. At the left, gas and air enter through flowmeters (not shown) and then into a proportioning mixer or carburetor. A compressor boosts the pressure of the mixture to force it through a firecheck and burner at the entrance to the reaction chamber, where it then passes through the catalyst bed, which is heated by the products of combustion and helps the reactions to completion in a small space. The products of combustion then pass through a gas cooler, to condense out most of the water and on to the furnace, unless some additional form of purification is necessary.

19.4.2 PURIFIED EXOTHERMIC GAS:

It is sometimes desirable to remove carbon dioxide and water from rich exothermic gas for sintering applications. Gas producers with accessories are used in such cases to make purified rich exothermic gas, free from carbon dioxide and water. In typical instances, the input ratio of air to natural gas is 6.5 or 7 to 1. Taking an average of 6.75:1, the composition is about 12.0% hydrogen, 9.0% carbon monoxide, 0.0% carbon dioxide, 0.2% methane, and 78.8% nitrogen. The dew point usually is -40° F(-40° C) or lower.

Carburizing of carbon-steel parts results in growth and possible blistering; decarburizing causes shrinkage—either condition can result in wide dimensional changes. Purified rich exothermic gas, being only mildly reactive and quite stable, creates no noticeable carburizing or decarburizing on iron or carbon-steel parts, and affords relative ease of dimensional control. It is assumed that there might be slight carburizing of iron parts or decarburizing of high-carbon-iron parts, but this is so nominal that it cannot be detected in photomicrographs, physical properties or performance characteristics. Accordingly, for practical purposes, the gas is assumed to be neutral over the entire range from no carbon to high carbon when sintering P/M parts. No adjustment, change, or control of the purified rich exothermic gas is necessary, other than to keep it carbon dioxide-free and dry, which conditions are assured by means of the auxiliary purifying equipment.

Of course, to obtain reliable results, uniform conditions must be held within the furnace. In a tight furnace, with normally-closed end doors, such as the box- and roller-hearth types, dew points of -20° to +25° F (-29° to -4° C) and 0.0 to 0.2% carbon dioxide are commonly maintained in production. Open-ended mesh-belt type furnaces require substantially higher flow rates of atmosphere gas and better control of room drafts, to achieve comparable results.

When sintering carbon-steel compacts, regardless of the type of atmosphere gas used, there generally is a loss in the amount of graphite from the original mix to the final amount of combined carbon. This is normally attributed to reactions of the graphite with oxygen in various forms within the compact, to create carbon dioxide. The extent of graphite loss depends upon such factors as the amount of air, moisture, and oxides, degree of activity of the grades of graphite and iron used, temperature, and time. It is common to start with about 1 ¼ to 2 times as much graphite in the green compacts as is desired in the form of combined carbon in the final sintered parts. When using purified rich exothermic gas, the final combined carbon is uniformly dispersed from center to surface of the parts.

Variations in composition of fuel gas supplied to the purified exothermic gas producer naturally result in variations in the output gas mixture of hydrogen, carbon monoxide, and nitrogen. There are, however, no variations in the carbon dioxide and water, since these objectionable constituents are consistently removed in the purifying system. Accordingly, variations in the input fuel-gas supply create no noticeable effect on the product or in the operation of the equipment. The fact that purified exothermic gas is always carbon dioxide-free and dry is, therefore, an important determining factor in its selection for critical applications. Also, this atmosphere gas has low flammability and thermal conductivity, which are desirable attributes.

In purified exothermic gas producers, sooting of the catalyst and consequent deterioration of the effluent product gas, requiring frequent ''burning out'' of the reaction chambers for restoration, is rarely encountered. Such producers commonly run months or even years continuously without the ''burning-out'' procedure.

Some users find it desirable to''burn-out'' the furnaces briefly each weekend or so, when using dry non-decarburizing atmospheres. This is done to remove accumulations of carbon, caused by distillation of lubricants from the compacts, even though substantial preheaters or burnoff chambers are utilized ahead of the high-heat chambers. Burning out is accomplished by cooling the furnace down to about 1300° F (704° C), opening the end doors, lighting the gas flowing out the ends, and then shutting off the gas supply. As the gas burns in, air enters the heating chamber and comes in contact with incandescent carbon deposits. This causes a temporary increase in temperature and cleans out the carbon more effectively than can be achieved by any other method.

Purified rich exothermic gas has a wide range of applications. Its most common uses are for sintering iron, iron-copper, carbon-steel, copper-steel, and for copper infiltrating iron and carbon-steel parts. It is also used for sintering bronze compacts, and in a few instances, brass compacts on trays under covers.

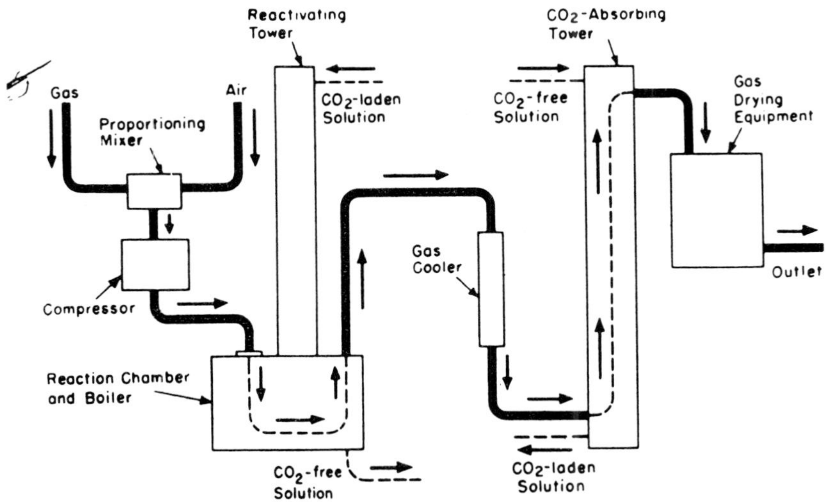

FIGURE 19D.
Simplified flow diagram of typical purified exothermic gas producer

A simplified flow diagram for a purified exothermic gas producer is shown in figure 19D. The initial stages are similar to those for the exothermic gas producer. Gas and air are metered, and pass through a proportioning mixture or carburetor, a compressor, a firecheck, and a burner, then into the reaction chamber. In this case, the reaction chamber is built into a boiler, so that the heat from the exothermic gas is given off to the carbon dioxide-laden chemical solution in order to boil the solution and help drive off the carbon dioxide. This is an economy measure as well, by utilizing heat that would otherwise be wasted.

The products of combustion pass on through a gas cooler and up through a carbon dioxide absorbing tower, through which is sprayed the carbon dioxide-free solution taken from the bottom of the reactivating tower. At this point, the gas leaves the producer and goes to drying equipment, which commonly consists of a refrigeration-type gas cooler to reduce the dew point to +40° F (4.4° C), and an activated-alumina dryer to further drop it to -40° F (-40° C) or lower. By thus scrubbing the undesirable constituents from the gas, the gas is always carbon dioxide-free and dry.

The relatively cool, carbon dioxide-laden solution from the bottom of the absorbing tower is passed through a heat exchanger, where it is heated by the boiling hot solution leaving the bottom of the reactivating tower. The hot carbon dioxide-laden solution is then charged into the top of the reactivating tower, is sprayed down over refractories where it gives up most of its carbon dioxide, and finally it runs down into the boiler. The solution, which is 10 to 15% monoethanolamine (MEA) in water, is constantly recirculated, and the operation is continuous and automatic.

Although the purified exothermic gas producer is rather complex at first glance, operators quickly become acquainted with its flow circuits and accessories, and find it to be a practical and dependable machine to manage. Routine checks of temperature and flows usually give adequate assurance of proper operation. About every eight hours the dual-chamber drier is switched over, a fresh drying tower is cut into the circuit, and the moisture-loaded tower is reactivated by heat to drive out accumulated moisture. Simple weekly tests are made for determining the effectiveness of the solution.

19.4.3 ENDOTHERMIC GAS:

In the operation of an exothermic gas producer, the minimum ratio of air to natural gas is about 6:1. At lower ratios the heat given off by the combustion reaction is insufficient to sustain the reaction. However, if an additional source of heat is used to supply the needs of the reaction and provide heat for the thermal losses of the chamber, the air-gas ratio can be reduced to any value desired. Such a reaction, which requires the addition of heat for its completion, is called "endothermic". Thus the products of reacting a hydrocarbon gas and air over a catalyst with heat added, are commonly known as "endothermic gas".

In figure 19B, the compositions of endothermic gas at various input air-gas

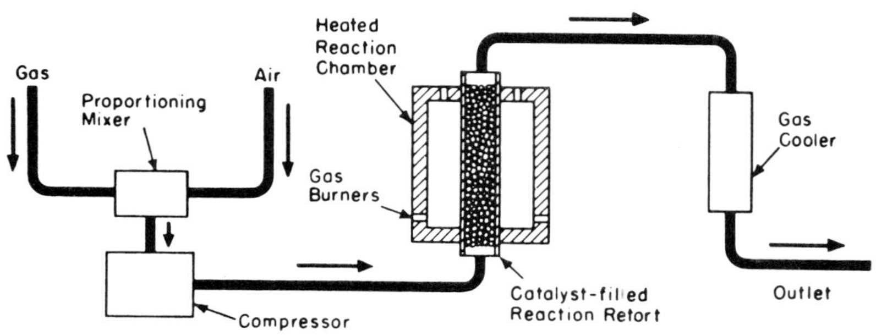

FIGURE 19E.

Simplified flow diagram of typical endothermic gas producer

ratios are shown. The maximum ratio at which some endothermic gas producers operate is 4.5:1 on vertical dash-line D. These units have silicon-carbide retorts. Vertical dash-line F is about the minimum ratio, 2.4:1, at which endothermic gas producers will function.

At the maximum ratio of 4.5:1, the composition of the endothermic gas is about 21.5% hydrogen, 13.8% carbon monoxide, 3% carbon dioxide, and 61.7% nitrogen. This gas is saturated with water at about 10° F (about 6° C) above the temperature of the cooling water used in the surface cooler of the producer. Figure 19B shows 80° F (27° C) dew point with 70° F (21° C) water, at high ratios. At 2.4:1 ratio, the composition is about 38% hydrogen, 20% carbon monoxide, 0.5% methane, 41.5% nitrogen. Due to a characteristic of the reaction, this gas is very dry, having a dew point of +10° to -10° F (-12° to -23°C), without the need of auxiliary drying equipment. This dry, high-carbon monoxide mixture has important and desirable properties. The particular thing to note at the moment is that at 4:5:1 and down to 2.75:1 ratio, the gas is wet, while at 2.4:1 it is dry.

The overall chemical reaction between natural gas (methane) and air, taking place in the endothermic gas producer, is represented by the equation:

$$2CH_4 + \text{air } (O_2 + 3.8N_2) = 2CO + 4H_2 + 3.8N_2$$

By using an adequate catalyst and a sufficiently high reaction temperature with an accurately-held air-gas ratio, the composition of the product gas can be controlled within desired limits. It will be noted in figure 19B that from 2.4:1 to about 2.9:1 ratios, the carbon dioxide content is approximately zero, but from 2.9:1 to 4.5:1 the carbon dioxide gradually creeps up to about 3%. Also, at 2.4:1 there is about 0.5% methane, but at about 2.9:1 the methane decreases to zero.

In figure 19B, it is seen also that the dew point drops off rapidly between the air-gas ratios of about 2.75:1, at which it is about 80° F (27° C), depending upon the cooling-water temperature and 2.4:1, at which it is about -10° F (-23° C). This range is useful for making gases with correct carbon potential. It is seen by the dotted dew point curve that minor changes in the ratio result in major changes in dew point. Also, it is known that the lower the water (and carbon dioxide), the higher the carbon potential, and vice versa. If the carbon potential of the gas is not in equilibrium with the carbon content of the metal, then the gas may either decarburize or carburize the work. The low density of most P/M parts makes them more sensitive to changes in carbon potential than solid steels, and undesirable effects can penetrate more deeply.

Changes in composition of the gas supplied to the generator can have the same over-all effect as changing the air-gas ratio. For best results it is necessary to hold a constant ratio and to use a fuel gas of constant composition. Gas producers are available which hold gas ratios after being set. Some natural gas supplies are constant, but others are subject to unannounced variations in composition. Best assurance of constant fuel-gas composition is sometimes obtained by resorting to use of propane made from natural gas, which is quite uniform in composition.

For critical jobs, dew points or carbon dioxide content of the atmosphere gas should be checked frequently. Any drift from normal can be corrected before trouble develops. Also automatic carbon-potential controllers, which regulate either dew point or carbon dioxide, are useful.

In sintering furnace applications, care must be taken to avoid contamination of the furnace-atmosphere samples by lubricants distilled out of the work. Typical distillation products are carbon and zinc. Such contaminants can upset the accuracy of the dew point controller. The critical zone of control is the high-heat chamber. Therefore an adequate auxiliary pre-heat or burn-off chamber is necessary to remove the lubricants before the work reaches the high-heat chamber. Also, ample flow of atmosphere gas through the high-heat chamber is needed to sweep any possible contaminants out through the end where the work enters.

It is apparent from figure 19B that wet endothermic gas at 3.5:1 ratio, vertical dash-line E, which is about the highest ratio used with alloy retorts in producers, is fairly high in reducing properties; it contains about 27.6% hydrogen and 16.5% carbon monoxide, totalling about 44.1% reducing constituents. This saturated gas, also containing a small amount of carbon dioxide, has no tendency to carburize low-carbon iron parts. Its highly-reducing property makes it quite useful in general P/M applications as well.

Because of its rather high hydrogen and carbon monoxide content, endothermic gas is quite flammable and must be handled with the same respect as hydrogen and dissociated ammonia. Dry endothermic gas is somewhat lighter than air, with a specific gravity of 0.622. Its high thermal conductivity, with respect to air, is about 3.23:1.

Applications of wet endothermic gas which has low carbon potential, include sintering compacts made of iron, iron-copper, copper, bronze, silver, nickel, etc., and infiltrating iron parts with molten copper. Uses of dry endothermic gas which can have medium or high carbon potential, include annealing of low-alloy steel powders with medium carbon; sintering compacts of carbon-steel, copper-steel, copper, bronze, brass, etc.; and infiltrating carbon-steel components. It is quite versatile and has a wide field of usefulness.

A simplified diagram of an endothermic gas producer is shown in figure 19E. As with other producers, hydrocarbon fuel gas and air are measured, mixed, and compressed, then passed into a reaction chamber. In this case, the catalyst is retained in a retort of silicon carbide or heat-resisting alloy. The retort is surrounded by the reaction chamber, heated with either gas burners or electric heating units, in the temperature range of about 1800° -2200° F (982° - 1204° C), depending upon the equipment and requirements. Following the reaction in the retort, the gas is cooled and goes on to the point of use.

19.5 ARGON AND HELIUM:

Argon (A) is available in steel cylinders. It has some use in P/M work because it is extremely inert. It can be used for sintering refractory and reactive metals (Table 19.2) at elevated temperatures.

Argon is non-flammable. It is somewhat heavier than air, its specific gravity being 1.379. Its thermal conductivity, 0.745, is lower than that for air.

Helium (He) is available in steel cylinders. It is equally as inert to the reactive metals as argon and, in general, is used interchangeably with argon for the same applications.

Helium is very light in weight, with a specific gravity of 0.137. It is the lightest of all non-flammable gases. It, like hydrogen, has high thermal conductivity, 6.217 compared to 1.0 for air.

19.6 VACUUM:

Up to this point, various atmosphere gases used for protecting different materials while being heated and cooled have been discussed. Some materials, however, are highly reactive and will combine with some of these gases or their impurities even though the gases are "purified". Such reactions may seriously impair some properties of the materials. In many such cases, however, it has been found that these materials can be treated in a vacuum with considerable success.

Vacuum furnaces produce an environment which retains the proper chemistry of the parts during sintering. With vacuum, reduction of oxides takes place while the parts are at temperature, and outgassing of the parts also is accomplished. In vacuum sintering, the environment is such that it is neither carburizing nor decarburizing during the sintering process.

Vacuum furnaces often are fitted with fibrous-graphite components (heating elements and insulation), figure 19F. The low-mass design results in fast-cycling furnaces with long component life, even at the tungsten-carbide and titanium-carbide sintering temperatures of approximately 2700° F (1482° C) (figures 19G and 19H).

Vacuum equipment is so flexible it can be adapted to the sintering and heat treating of alloyed components, e.g., first sintering, then precipitation-hardening, solution annealing (including oil quenching), and aging. All of this can be accomplished without unloading the furnace until the sequence is finished.

Considerable production and experimental sintering is being done in vacuum. Most of this work is on the reactive metals which are highly susceptible to the formation of hydrides, nitrides, or oxides in gaseous atmospheres. Good results are obtained by vacuum sintering refractory metal carbides, stainless steels, beryllium, titanium, zirconium, tantalum, niobium (columbium), vanadium, thorium, uranium, and metal-ceramic combinations (cermets).

FIGURE 19F.

Vacuum sintering furnace with fibrous-graphite components, showing racks of stainless-steel compacts

FIGURE 19G.

Front view of dual chamber vacuum furnace for sintering high temperature materials such as carbides

FIGURE 19H.

Rear view of dual chamber vacuum sintering furnace shown in Figure 19G

Protective atmospheres in furnaces normally are used at atmospheric pressure, which is about 760 mm mercury. Most high-vacuum work is done at pressures considerably lower than 1 mm Hg (1 Torr), so vacuum pressures often are expressed in terms of microns of Hg. One micron is 0.001 mm or 1×10^{-3}mm. One tenth micron is 1×10^{-4}mm, 1/100 micron is 1×10^{-5}mm, and so on.

The vacuum pumping system is directly connected to the chamber in which the work is heated and cooled. The chamber usually is pumped down to the desired pressure before heat is applied. During heating, however, there may be bursts of absorbed gases from the work. These will increase the chamber pressure and may require that the temperature be held constant until the pressure is pumped down again to the desired level. Organic materials in pressed powder compacts are particularly prone to violent outgassing due to decomposition. It is not uncommon to use a moderate temperature presintering treatment in a pure gas atmosphere to drive off the volatiles and avoid overloading and/or contaminating the vacuum system. Even then, when the compacts are subsequently sintered in vacuum, it may be necessary to install a "cold trap" ahead of the vacuum pump to collect condensable vapors, or to use one of several other expedients to remove them. This will assure best pumping efficiency and minimize contamination of the vacuum-pump oil.

Vacuum often is more economical than using protective atmosphere gases, particularly bottled gas. The only operating costs involved in producing the vacuum are electrical energy and oil for the pumps. Vacuum furnaces, however, often are more expensive than equivalent protective-atmosphere furnaces for the same work, due to the necessity for vacuum-tight welding, leak testing, and strengthening retorts and bases against high collapsing pressures.

Vacuum pumps commonly used are mechanical pumps and oil-vapor pumps. Two pumps of similar or unlike types are often used in tandem for powder metallurgy work, one (the mechanical "forepump"), having a dual purpose as "roughing and backing pump" roughly to reduce the pressure to a value, at which point the vapor stream or diffusion pump can be cut in to further reduce the pressure.

Rotary, positive-displacement, mechanical pumps generally fall into three subdivisions: (1) single stage oil-sealed pumps for use from 760 Torr to 10 microns Hg McLeod, (2) compound oil-sealed pumps for use from 760 Torr to 0.2 microns Hg McLeod, and (3) oilless mechanical booster lobe pumps for use from approximately 50 Torr to below 0.1 micron Hg McLeod. This last group (3) must be operated in tandem with a standard rotary oil-sealed pump, as it can not be run at atmospheric pressure except in very small sizes.

Diffusion-ejector pumps and diffusion pumps with mechanical backing pumps handle large throughputs between the range of about 650 and 2 microns, and 10 to 0.1 microns Hg, respectively. Oil diffusion pumps with mechanical backing pumps are effective from 100 microns down to 0.0001 microns (1×10^{-7} Torr). The actual pressure ranges vary somewhat with size and make. The ranges given here are typical.

20.0 INSTRUMENTS FOR ANALYZING AND CONTROLLING GASES

There are a wide variety of instruments for analyzing and controlling gases used as protective atmospheres for powder metallurgy. A few typical ones will be discussed to illustrate the convenience and helpfulness of having such equipment available, and to describe the general principles on which they operate.

20.1 ORSAT-TYPE ANALYZERS:

Complete analysis of a gaseous mixture generally is divided into two parts: first, the absorption phase; and second, the explosion or burning phase. Percentages of carbon dioxide, oxygen, and carbon monoxide are first determined by chemical absorption in the order named. This is generally called the Orsat principle. Orsat analyzers can be obtained for measuring carbon dioxide, oxygen and carbon monoxide, or for only carbon dioxide or oxygen.

Such units are portable and permit relatively rapid analyses. Hydrogen and hydrocarbons such as methane are measured by explosion or slow burning, in a unit such as the Morehead burette, which will not be described here because of its infrequent use in powder metallurgy work.

In an Orsat analyzer, a sample of the gas mixture is trapped in a calibrated glass water-jacketed chamber, and the sample is accurately measured. The sample is then passed through an absorbing medium in one of three glass absorption pipettes where carbon dioxide is removed from the mixture. The gas sample is then returned to the measuring chamber where its volume is again indicated. The difference between the volumes before and after the absorption gives the percentage of carbon dioxide absorbed by the medium. In similar fashion, the oxygen and carbon monoxide contents are determined in consecutive order in the other absorbing pipettes. Manipulation of the gases from one chamber to the other is accomplished by raising and lowering a water-leveling bottle and turning cocks in the gas connecting lines.

Another unit which operates on the Orsat absorption principle is available for measuring only carbon dioxide or oxygen, quickly and simply. It consists of a small calibrated plastic tube with a reservoir at top and bottom, the unit being of "hour glass" shape. A gas sample is pumped in to fill the upper reservoir, while the absorbing liquid is in the bottom reservoir. The analyzer is then turned upside down, and the liquid flows down through the absorption tube. Then the unit is turned upright again for repeating the absorption cycle. Absorption of gas results in a partial vacuum, causing the column of liquid, supported on a diaphragm, to rise in the tube. Any change in height of the liquid column, when comparing readings taken before and after the manipulations, indicates by direct percentage reading the carbon dioxide or oxygen content of the gas sample.

20.2 SPECIFIC-GRAVITY ANALYZERS:

The specific gravity of gases being tested can be measured and compared against that of air. Since carbon dioxide is much heavier than the other constituents in such mixtures, an instrument of this type is especially sensitive to changes in carbon dioxide. Therefore the instrument is useful, for example, in checking sintering-furnace atmospheres of purified rich exothermic or rich endothermic gases which are supposed to be carbon dioxide-free and dry. Should carbon dioxide appear in the influent or effluent furnace-atmosphere gases, a pronounced indication would result and show the need for corrective measures.

As another example, the inert gas used for purging air from a furnace at start-up, or flammable gas during shutdown and the flammable gases used for the atmosphere, are appreciably lighter (or heavier) than air. By continuously measuring the specific gravity of the effluent gas from the furnace, the completeness of purge can be determined. This is important, to prevent explosions and to avoid contaminating the atmosphere by air or moisture which would form in an incompletely purged furnace.

For automatically operating mixing or regulating valves, or other control mechanisms, a pneumatic transmitter can be installed in the instrument. The instrument also can be equipped with electronic contactors for signalling faulty conditions or, with interlocks, to prevent startup or create a shutdown if hazards exist.

The analyzer contains two motor-driven impellers rotating in opposite directions and located in separate chambers. Facing each impeller is a companion impulse wheel. The shafts for these impulse wheels are tied together by external linkage. The action of each impeller and impulse wheel is somewhat similar to the fluid drive in automobiles except that the gas and air are substituted for the oil.

A sample of gas to be tested is drawn into one chamber by an impeller; the rotation creates a torque on the impulse wheel which is proportional to the gas density. Atmospheric air is drawn into the other chamber, and in like manner results in a torque in the opposite direction on the other impulse wheel. The differences between these opposing torques in the impulse wheels is a measure of the specific gravity, and results in the movement of a pointer on a scale. A record can also be made on a chart and the signal can be used for operating some sort of control device.

20.3 INFRARED ANALYZERS:

Different gases absorb infrared energy at charactertistic wave lengths. Because of this property, changes in the concentration of a single component in a mixture produce corresponding changes in the total energy remaining in an infrared beam passed through the mixture. These energy changes, which are detected by an infrared analyzer, are therefore a measure of the gas concentration. Proper selection of apparatus permits accurate, rapid analyses for such constituents as carbon monoxide, carbon dioxide, and methane. One make of instrument also measures water in terms of dew point. Each gas compound absorbs a certain portion of the infrared spectrum which no other gas absorbs, and the amount of radiation absorbed is proportional to the concentration of the specific gas. Such instruments are not suitable for measuring gases having no infrared absorption band, such as oxygen, hydrogen, and nitrogen.

Infrared analyzers have very high sensitivity and selectivity. An example is the measurement of 0 to 20 parts per million carbon dioxide in a 6-component hydrocarbon stream.

Typical applications of infrared analyzers are in the field of gases with high carbon potential, such as purified rich exothermic gas or dry endothermic gas, and those with high purity such as dry hydrogen or argon. Either water or carbon dioxide is a good indicator of carbon potential. Small changes in the amounts of these constituents, in some situations, can make wide changes in carbon potential. Also, differences in dew point of dry gases such as hydrogen or argon can affect sintering results on reactive metals. The high sensitivity of the infrared analyzer is of value, particularly for a large furnace with high output, where the substantial cost of this type analyzer can be justified. These analyzers also can be equipped to control the level of water or carbon dioxide to provide carbon-potential control.

20.4 MOISTURE DETECTORS:

The moisture content of an atmosphere gas often has a direct bearing on results obtained in sintering. The most convenient means of determining moisture content is by checking the dew point. Sometimes it is necessary or desirable to know the dew point of a gas entering and leaving a refrigeration-type gas cooler or regenerative-type dryer. The dew point of a gas sample taken from a furnace chamber is frequently needed, in comparison with entering dew point, to determine the increase of water, if any. Excess water in a supposedly non-decarburizing atmosphere can cause decarburization; or expressed in another way, the water content usually is a good indicator of the carbon potential. Since carbon dioxide reacts with hydrogen at elevated temperatures to form water and carbon monoxide, by the water-gas reaction, an increase in the dew point of supposedly carbon dioxide-free dry atmosphere containing hydrogen will indicate the presence of carbon dioxide in that atmosphere as being the real source of the water. Also, water often causes oxidation of reactive metals, even when present in relatively small quantities in otherwise pure atmospheres. Therefore, the dew-point determination is a good all-round method of making a quick check to see whether conditions are right throughout a furnace system.

A relatively simple and inexpensive apparatus for indicating dew point is called a "dew cup". This device consists of an outer container with an observation window, through which the gas passes to be tested. Within the container is a cup, nickel-plated and highly polished on the outside. This cup is situated directly in the gas stream so that the entering gas will impinge directly on its polished surface. Within the cup is a glass thermometer. To make a test, the atmosphere gas is directed to flow through the container, and acetone is poured into the cup. After about 5 minutes, small amounts of crushed dry ice are added to the acetone, while stirring constantly with the thermometer. At the first sign of dew or moisture on the polished surface of the cup, the temperature is read from the thermometer. This reading is the dew point of the gas being tested.

Another type of dew point indicator, somewhat more complex but actually relatively simple to operate, is an instrument which involves compressing a sample of the gas, then quickly expanding it. If the gas, due to its rapid expansion, has cooled below its dew point, a fog will be observed in the expansion chamber. The procedure is repeated to find the end point or vanishing point of the fog or condensation. A reading is taken from a pressure-ratio gauge, which indicates directly the relation between the pressure of the gas sample at the end point and of the atmospheric pressure. By means of a separate dial calculator, this pressure ratio is converted to dew point in degrees F.

The amount of water vapor in an atmosphere gas can be determined by its action on a hygroscopic salt. If dry lithium chloride salt, for example, is exposed to the atmosphere gas, it will absorb moisture and dissolve, forming a saturated solution. If the salt and the saturated solution are heated, the water in the solution will evaporate. At equilibrium temperature the vapor pressure of the water in the solution is equal to the partial pressure of the water vapor in the atmosphere gas. This temperature is a measure of the partial pressure of water vapor in the atmosphere gas—that is, the dew point temperature.

The humidity-sensitive element is a thermometer bulb inside a thin-walled metal tube covered with a woven-glass tape impregnated with lithium chloride. The tube is wound with a pair of gold-overlay wires over the tape, and is surrounded by a perforated metal guard. When the salt absorbs moisture from the gas, it becomes an electrical conductor. Current is passed between the two wires to raise the temperature of the element until equilibrium is attained. This temperature is sensed by the thermometer bulb inside the tube and transmitted to a recorder calibrated to read directly the dew point of the gas.

Application of this method, which is useful for indicating, recording, and even controlling dew points, can be made in connection with sintering furnaces in which carbon potential, hence dew point, is critical. The system is particularly useful for gases entering the furnace. Control is achieved by automatically adjusting the ratio of the air and gas entering the gas producer. If contaminants, such as distillation products from lubricants in the work, are negligible in the effluent gas from the furnace, the system can be extended further to control the effluent dew point. This is quite desirable, since effluent gas is much more representative than influent gas of actual conditions surrounding the work. In this case, fastest response generally is achieved by injecting and regulating a small amount of hydrocarbon gas, such as methane or propane, directly into the heating chamber. The dew point controller automatically regulates a valve in the auxiliary gas line to hold a constant pre-set dew point within the furnace, and thereby maintain uniform carbon potential.

Another useful type of dew point indicator and recorder is the infrared type of one particular make, as discussed under "Infrared Analyzers".

20.5 CARBON POTENTIAL CONTROL:

The carbon potential of a furnace atmosphere may be defined as its carbon equilibrium with steel at a specific temperature. For example, to sinter an iron-graphite mixture and maintain an 0.80% carbon in the resulting steel at 2050° F (1121° C), the furnace must have a carbon potential of 0.80% C at the sintering temperature of 2050° F (1121° C). As mentioned in the discussion on instrumentation, either dew point (water vapor) or carbon dioxide content can be used as a measure of carbon potential of an endothermic-type sintering atmosphere. To put into successful practice a carbon-potential control system, one must understand the following: (a) sampling, (b) measurement, (c) control principles, and (d) effect of various temperature zones on control.

Sampling: The gas sample taken for measurement of the carbon potential should be taken from the high-heat, sintering soak zone of the furnace. A sample tube of a heat resisting alloy or a refractory tube of ¼ in. I.D. will serve this purpose. For sintering temperatures up to 2100° F (1149° C), a heat-resisting alloy tube, such as Inconel, is more convenient to install gas-tight than a refractory tube. If the furnace is of a non-muffle type, the tube must extend at least two inches into the hot zone beyond the refractory wall. If the furnace is of the muffle type, the sample tube must be welded to the muffle and clearance allowed in the refractory brick and shell of the furnace for movement due to expansion of the muffle. It is very important to connect a tee at the end of the sample tube that extends outside the furnace shell, so that a valve or clean-out plug can be connected for daily cleaning with a wire brush, or by blowing compressed nitrogen through the tube to remove carbon deposits. Carbon deposits in the sampling tube will cause erroneous dew point or carbon dioxide readings. Copper or aluminum tubing of ⅜ in. O.D. size can be connected to the 90° outlet of the tee on the gas sample tube. A filter should be installed about 15 inches from the outlet of the gas sample tube to prevent the extension tube from becoming contaminated between the furnace and the instrument. Another filter must be installed immediately ahead of the instrument to assure that no dirt or condensation enters the measuring instrument.

Since the furnace is not under positive pressure, a gas sample pump will be required to pump the sample from the furnace through the instrument. Since most instruments are flow-sensitive, it is essential to follow the instrument manufacturer's recommendations on the flow requirements. A small flowmeter usually is used to measure the required gas sample, and to maintain consistent results from the instrument. The endothermic generator operates under pressure, and thus a sample pump is not required for it. The same recommendations of double filters, however, and a sample measuring flowmeter apply.

Measurement: Instruments to measure a constituent to control the carbon potential have been described above. These instruments are for either dew point(moisture control)or carbon dioxide measured by infrared. The instruments can be indicating with a manual control, or indicating or recording with full automatic control. The choice of the dew point type or the infrared carbon dioxide type is a matter of economics and application. The dew point instruments are less costly than the carbon dioxide infrared analyzers. For single-point control, both will do a satisfactory job. Dew point analyzers are not as sensitive, and are slower in response than carbon dioxide infrared analyzers. This may or may not be a disadvantage. Because of the fast response, infrared analyzers are used for multi-point control systems. For example, when one analyzer is used to measure and control the generator, the sintering furnace, and the hardening furnace, the infrared carbon-dioxide analyzer is used because of the speed of response and sensitivity of measurement. Automatic diverting valves are used to switch the sample gas from the generator to the sintering furnace and to the hardening furnace at one-minute intervals. Thirty seconds are allowed to purge the instrument with the new process gas sample, and thirty seconds are then allowed for the instrument to analyze the gas and to send the signal to the controller to make the proper adjustment to bring the process under control. When infrared analyzers are used for multi-point control, the cost then can compete per point of control with dew-point analyzers. Dew point is not satisfactory to use for multi-point control because water molecules are difficult to purge from an instrument in a short time.

Control: Whether the process is to be controlled manually or automatically, the same principle applies. For carbon potential control of an endothermic-type atmosphere, both the generator and the furnace should be controlled to achieve the best results. If the process is either sintering or hardening of P/M parts where it is desired to control the carbon on the part above the 0.50% carbon range, the generator should be controlled at 25° - 30° F (-3.9° to -1.1°C) dew point or 0.25 to 0.30% carbon dioxide.

To control the process, whether it be sintering or hardening, it is necessary to refer to equilibrium data of dew point vs. carbon content, or carbon dioxide vs. carbon content, at the operating temperature of the process. Suppose one desired to sinter an iron-1% graphite mixture at 2050° F (1121° C) and maintain an 0.80% carbon content in the resulting sintered steel: referring to figure 17A, the equilibrium relationships between dew point and percent carbon in plain carbon steels at various furnace temperatures, shows that a 15° F(-3.4° C) dew point is required. Since the endothermic gas entering the furnace will be between 25° - 30° F(-3.9° to -1.1° C), the dew point will be much higher than shown by the equilibrium data. To achieve the 15° F (-3.4° C) dew point in the furnace, natural gas or another hydrocarbon gas such as propane, is added to the endothermic gas between the generator and the furnace. The mixture is put into the hot sintering zone only, and at several inlets. The hydrocarbon gas reacts with the water vapor and carbon dioxide constituents in the endothermic gas in the furnace at 2050° F (1121° C) to reduce the decarburizing constituents to the desirable level, as shown by the equilibrium data. If the process is controlled manually, the hydrocarbon gas is added in small increments by a needle valve and measured by a small precision flowmeter. If the process is automatic, the control instrument sends a signal to a motorized valve or pneumatic valve to add the proper amount of hydrocarbon gas to achieve the desired control point. It is very important that the gas sample tube be not adjacent to the gas inlet tube. The endothermic gas and the hydrocarbon gas must be fully reacted before being withdrawn from the furnace to be analyzed by the instrument.

If the process to be controlled is to sinter low-carbon iron parts requiring dew points higher than the 25° - 30° F (-3.9° to -1.1° C) generator dew point, this can be achieved either by controlling the generator output at a higher dew point, or by adding air to the endothermic atmosphere instead of a hydrocarbon gas. In some instances where both high and low dew points are to be controlled in the furnace, automatic controllers are designed to add either air or hydrocarbon gas, as the process dictates.

The equilibrium data in figure 17A were determined by actual measurement, and are fairly accurate. However, the data should be used only as a guide to tune the process in to control. The final adjustment should be made after the carbon analysis of the P/M sintered part is determined. In view of the fact that carbon dioxide measurement of carbon potential is newer than dew point, accurate carbon dioxide vs. carbon equilibrium data are not available. When controlling carbon dioxide, it is recommended that a small portable indicating dew point meter be used in conjunction with figure 17A to tune in the process. Then the corresponding carbon dioxide reading on the infrared unit should be used as the control point. One can then make his own carbon dioxide vs. carbon content equilibrium diagram for his own particular furnace.

Effect of Various Temperature Zones on Control: Referring to figures 16A and 16B, there are various temperature zones in a continuous sintering furnace. To achieve a specific carbon content in the sintered product, it is important to understand the effect of temperature of each zone on the final carbon equilibrium of the P/M part. The burn-off or preheat section, and possibly the first zone of the high-heat section, will run at a lower temperature than the final high-heat sintering zone. By referring to the equilibrium diagram in figure 17A, one will note that if the carbon potential of the atmosphere is adjusted for the 2050° F (1121° C) high-heat zone, an excessive amount of carbon will be picked up at the lower temperatures. This is no problem in reference to the final desired carbon in the P/M part, as the excessive carbon will be extracted in the high-heat zone to the proper equilibrium point. However, any lower-temperature zones following the high-heat zone will add carbon, as shown by the equilibrium diagram. Thus excessive carbon over the desired amount will remain after cooling.

This is of no great concern when one wishes to sinter and maintain the maximum carbon content of 0.80% to 1.00%. Generally speaking, it is difficult to maintain these levels even with the best of sintering furnaces and atmosphere controls, and the build-up of carbon in the insulated cooling zone will not exceed the desired high carbon content for the short time the parts are exposed to the high carbon potential atmosphere at the lower temperature.

It may be a problem, however, to control the carbon at some exact equilibrium point under the maximum of 0.80 - 1.00%, say 0.30 or 0.50%, to overcome the effect of the varying lower-temperature range of an insulated precooling zone. Several things may be done to achieve a semblance of control. First, eliminate the insulated cooling zone so the parts will cool rapidly as they enter the water-jacketed cooling zone. This may prevent sufficient time in the transition-temperature range for carbon pick-up. Eliminating the insulated cooling zone, however, may cause higher maintenance on the first water-jacketed chamber, and may also cause too much thermal shock on the P/M parts being sintered.

The second solution is to dilute the endothermic gas in the transition zone with a gas that has no carbon potential, such as dissociated ammonia or nitrogen. For example, two-thirds of the total atmosphere-gas flow can be enriched endothermic gas to achieve the desired carbon potential put into the high-temperature sintering zone, and one-third dissociated ammonia can be put into the intermediate-temperature insulated pre-cooling zone. By trial and error, the exact proportions can be worked out to achieve the desired surface carbon chemistry.

Another solution is to put a sample tube into the insulated pre-cooling zone and add another control point to the instrument. A lean, unpurified, exothermic gas can be added to the unenriched endothermic gas supplied to this zone, if necessary, by the controller to achieve a dew point or carbon dioxide level to be in approximate equilibrium with the P/M parts in the varying-temperature transition zone. The water vapor or carbon dioxide set point for control will have to be worked out by trial and error. Once the point is determined for a specific production rate, automatic control can be achieved from day-to-day.

20.6 HIGH-VACUUM GAUGES:

It is very important in vacuum powder metallurgy to know the pressure in heating chambers and perhaps in headers, manifolds, and at the pumps. A variety of vacuum gauges are available. Some of the more common ones will be mentioned here to acquaint you with their names, operating principles, and pressure ranges.

The McLeod gauge is a mercury manometer of such precision that it is often used for checking the calibration of other types of vacuum measuring and control instruments. By manipulating the McLeod gauge and measuring the difference in height of two mercury columns in adjacent capillary tubes, it is possible to read the pressure of the system in microns mercury on a calibrated scale. In the manipulation, the measured sample is compressed to approximately atmospheric pressure by the column of mercury in one deadend capillary. Any vapors, such as moisture, will condense under this pressure and introduce an error. The gauge gives true readings only with dry permanent gases such as nitrogen, oxygen, and air. McLeod gauges are built to cover various ranges. One covers from about 5.0 mm (5.0 Torr or 5000 microns) to 1×10^{-5} mm (0.01 micron) Hg; another about 2 microns to 1×10^{-7} mm Hg.

The Pirani gauge employs a Wheatstone bridge circuit to balance the resistance of a resistor or of a tungsten filament, sealed off in high vacuum against that of a tungsten filament which can lose heat by conduction to the gas whose pressure is being measured. A change of pressure causes a change in the filament temperature, and, consequently, of the wire resistance, thus unbalancing the bridge. The amount of this unbalance is indicated on a microammeter and read in terms of pressure. Typical Pirani gauges cover a range of about 2.0 mm (2.0 Torr) to 0.001 mm (1 micron) Hg; but one type covers the unusually broad operating range of about 100 Torr to 0.001 Torr (1 micron Hg).

The thermocouple gauge consists of a thermocouple connected to a tiny heating unit in a sensing tube placed in the vacuum system. A constant current through the heater produces a given temperature, resulting in an EMF (electromotive force) across the terminals of the thermocouple which is fed through a calibrating potentiometer to a simple meter for pressure indication. As the pressure is reduced, so is the cooling effect of the surrounding gas, resulting in an increase in the temperature of the thermocouple. Such variations in temperature are thus used to indicate variations in pressure in the system. Thermocouple vacuum gauges have ranges of about 1.0 Torr to 0.001 Torr (1 micron Hg).

Ionization gauges are of two types: cold-cathode (ordinarily called "Philips" or "discharge" gauges), and hot-filament (usually simply called "ionization" gauges). The principle of operation is that collisions between molecules of gases and electrons result in formation of ions. Below about 1 micron Hg this formation of ions varies linearly with pressure. Measurement of this ion current can be translated into units of gas pressure.

In the cold-cathode gauge, the ion current is produced by a high-voltage discharge. The cathode and anode are suspended in a magnetic field in the sensing element. The electrons emanating from the cathode are caused to spiral as they move across the magnetic field to the anode and they ionize gas molecules in their paths. The ionization current is directly proportional to the molecular density, and when passed through a microammeter it gives a pressure indication. The Philips gauge has a wide operating range, a typical one being from 25 microns to 1×10^{-5} Torr (0.01 micron Hg).

In the hot-filament ionization gauge, the electrons emitted from a heated filament in the sensing tube are accelerated toward a positively-charged cylindrical grid. Some electrons pass into the space between the grid and a negatively-charged collector and collide with gas molecules from the vacuum system to produce positive ions. Here again the ion current is proportional to the number of molecules and to the pressure which is read on a microammeter. Hot-filament gauges have several ranges, all starting with a maximum of 1 micron Hg, with typical minimums of about 1×10^{-5}, and 2×10^{-12} Torr.

Another type of ionization gauge is the Alphatron gauge which utilizes a radioactive source to produce alpha particles that ionize the gas molecules. This radioactive source replaces the filament, or electron source, in the hot-filament gauge. The range of the Alphatron gauge is wide—in one typical case from 1000 Torr to 0.0001 Torr (0.1 micron Hg).

To summarize, the McLeod, Pirani, and thermocouple gauges are used down to pressures of about 1 micron Hg, the Alphatron gauge to 0.1 micron Hg, and the ionization gauges starting with a top pressure of 1 micron and going down to extremely low pressures, in one case to 2×10^{-12} Torr. The actual pressure ranges vary somewhat with type and make, and the ranges given here are just typical.

— Notes —

— Notes —

— Notes —

— Notes —

— Notes —

— Notes —

— Notes —

— Notes —

— Notes —

— Notes —

— Notes —

— Notes —

— Notes —

— Notes —

— Notes —

— Notes —